Learning Algorithms

A Programmer's Guide to Writing Better Code

George T. Heineman

Beijing · Boston · Farnham · Sebastopol · Tokyo

Learning Algorithms

by George T. Heineman

Published by O'Reilly Media, Inc., 1005 Gravenstein Highway North, Sebastopol, CA 95472.

O'Reilly books may be purchased for educational, business, or sales promotional use. Online editions are also available for most titles (*http://oreilly.com*). For more information, contact our corporate/institutional sales department: 800-998-9938 or *corporate@oreilly.com*.

Acquisitions Editor: Melissa Duffield	**Indexer:** nSight, Inc.
Developmental Editor: Sarah Grey	**Interior Designer:** David Futato
Production Editor: Beth Kelly	**Cover Designer:** Karen Montgomery
Copyeditor: Piper Editorial Consulting, LLC	**Illustrator:** Kate Dullea
Proofreader: Justin Billing	

August 2021:   First Edition

Revision History for the First Edition
2021-07-20:   First Release
2021-10-15:   Second Release

See *http://oreilly.com/catalog/errata.csp?isbn=9781492091066* for release details.

978-1-492-09106-6

[LSI]

Table of Contents

Foreword

Algorithms are at the heart of computer science and essential to the modern information age. They power the search engines used to answer billions of daily Internet search requests and provide privacy when communicating over the Internet. Algorithms are increasingly visible to consumers in countless areas, from customized advertising to online price quotes, and the news media is full of discussions about what algorithms are and what they can do.

The large growth in STEM (Science, Technology, Engineering and Mathematics) is powering a new wave of sustained growth and innovation in the global economy. But there simply aren't enough computer scientists to discover and apply the algorithms needed for advances in medicine, engineering, and even government. We need to increase the number of people who know how to apply algorithms to the problems within their own fields and disciplines.

You don't need a four-year degree in computer science to get started with algorithms. Unfortunately, most online material and textbooks on the topic are designed for undergraduate students, with an emphasis on mathematical proofs and computer science concepts. Algorithm textbooks can be intimidating because they are references for so many different algorithms, with countless variations and highly specialized cases. All too often, readers find it hard to complete the first chapter of these books. Using them can be a bit like trying to improve your English spelling by reading an entire dictionary: you would be much better off if, instead, you had a specially designed reference book that summarizes the 100 most misspelled words in the English language and explains the rules (and exceptions) that govern them. Similarly, people from different backgrounds and experiences who use algorithms in their work need a reference book that is more focused and designed for their needs.

Learning Algorithms provides an approachable introduction to a range of algorithms that you can immediately use to improve the efficiency of your code. All algorithms are presented in Python, one of the most popular and user-friendly programming languages, used in disciplines ranging from data science to bioinformatics to engineering. The text carefully explains each algorithm, with numerous images to help readers grasp the essential concepts. The code is open source and freely available from the book's repository.

Learning Algorithms will teach you the fundamental algorithms and data types used in computer science, so that you can write more efficient programs. If you are looking for a technical job that requires programming skills, this book might help you ace your next coding interview. I hope it inspires you to continue your journey in learning algorithms.

— Zvi Galil
Dean of Computing Emeritus
Frederick G. Storey Chair in Computing
Georgia Institute of Technology
Atlanta, May 2021

Preface

Who This Book Is For

If you are reading this book, I assume you already have a working knowledge of a programming language, such as Python. If you have never programmed before, I encourage you to first learn a programming language and then come back! I use Python in this book because it is accessible to programmers and nonprogrammers alike.

Algorithms are designed to solve common problems that arise frequently in software applications. When teaching algorithms to undergraduate students, I try to bridge the gap between the students' background knowledge and the algorithm concepts I'm teaching. Many textbooks have carefully written—but always too brief—explanations. Without having a guide to explain how to navigate this material, students are often unable to learn algorithms on their own.

In one paragraph and in Figure P-1, let me show you my goal for the book. I introduce a number of data structures that explain how to organize information using primitive fixed-size types, such as 32-bit integer values or 64-bit floating point values. Some algorithms, such as Binary Array Search, work directly on data structures. More complicated algorithms, especially graph algorithms, rely on a number of fundamental abstract data types, which I introduce as needed, such as *stacks* or *priority queues*. These data types provide fundamental operations that can be efficiently implemented by choosing the right data structure. By the end of this book, you will understand how the various algorithms achieve their performance. For these algorithms, I will either show full implementations in Python or refer you to third-party Python packages that provide efficient implementation.

If you review the associated code resources provided with the book, you will see that for each chapter there is a book.py Python file that can be executed to reproduce all tables within the book. As they say in the business, "your mileage may vary," but the overall trends will still appear.

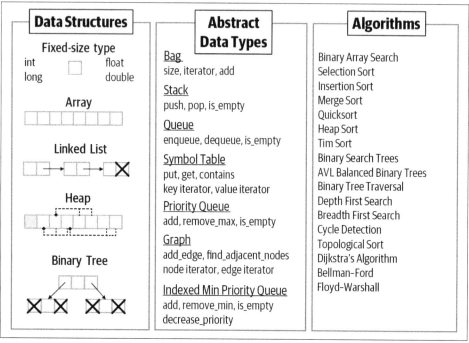

Figure P-1. Summary of the technical content of the book

At the end of every chapter in the book are challenge exercises that give you a chance to put your new knowledge to the test. I encourage you to try these on your own before you review my sample solutions, found in the code repository for the book.

About the Code

All the code for this book can be found in the associated GitHub repository, *http://github.com/heineman/LearningAlgorithms*. The code conforms to Python 3.4 or higher. Where relevant, I conform to Python best practices using double underscore methods, such as __str()__ and __len()__. Throughout the code examples in the book, I use two-space indentation to reduce the width of the code on the printed page; the code repository uses standard four-space indentation. In a few code listings, I format code using an abbreviated one-line if statement like if j == lo: break.

The code uses three externally available, open source Python libraries:

- NumPy (*https://www.numpy.org*) version 1.19.5
- SciPy (*https://www.scipy.org*) version 1.6.0
- NetworkX (*https://networkx.org*) version 2.5

NumPy and SciPy are among the most commonly used open source Python libraries and have a wide audience. I use these libraries to analyze empirical runtime performance. NetworkX provides a wide range of efficient algorithms for working with graphs, as covered in Chapter 7; it also provides a ready-to-use graph data type implementation. Using these libraries ensures that I do not unnecessarily reinvent the wheel. If you do not have these libraries installed, you will still be fine since I provide workarounds.

All timing results presented in the book use the `timeit` module using repeated runs of a code snippet. Often the code snippet is run a repeated number of times to ensure it can be accurately measured. After a number of runs, the minimum time is used as the timing performance, *not* the average of all runs. This is commonly considered to be the most effective way to produce an accurate timing measurement because averaging a number of runs can skew timing results when some performance runs are affected by external factors, such as other executing tasks from the operating system.

When the performance of an algorithm is highly sensitive to its input (such as Insertion Sort in Chapter 5), I will clearly state that I am taking the average over all performance runs.

The code repository contains over 10,000 lines of Python code, with scripts to execute all test cases and compute the tables presented in the book; many of the charts and graphs can also be reproduced. The code is documented using Python docstring conventions, and code coverage is 95%, using *https://coverage.readthedocs.io*.

If you have a technical question or a problem using the code examples, please send email to *bookquestions@oreilly.com*.

This book is here to help you get your job done. In general, if example code is offered with this book, you may use it in your programs and documentation. You do not need to contact us for permission unless you're reproducing a significant portion of the code. For example, writing a program that uses several chunks of code from this book does not require permission. Selling or distributing examples from O'Reilly books does require permission. Answering a question by citing this book and quoting example code does not require permission. Incorporating a significant amount of example code from this book into your product's documentation does require permission.

We appreciate, but generally do not require, attribution. An attribution usually includes the title, author, publisher, and ISBN. For example: "*Learning Algorithms: A Programmer's Guide to Writing Better Code* by George T. Heineman (O'Reilly). Copyright 2021 George T. Heineman, 978-1-492-09106-6."

If you feel your use of code examples falls outside fair use or the permission given above, feel free to contact us at *permissions@oreilly.com*.

Conventions Used in This Book

The following typographical conventions are used in this book:

Italic

Indicates new terms, URLs, filenames, file extensions, and points I want to emphasize.

`Constant width`

Used for program listings as well as within paragraphs to refer to program elements such as variable or function names, data types, statements, and keywords.

This element, identified by an image of a ring-tailed lemur, is a tip or suggestion. I use this image because lemurs have a combined visual field of up to 280°, which is a wider visual field than anthropoid primates (such as humans). When you see this tip icon, I am literally asking you to open your eyes wider to learn a new fact or Python capability.

This element, identified by an image of a crow, signifies a general note. Numerous researchers have identified crows to be intelligent, problem-solving animals—some even use tools. I use these notes to define a new term or call your attention to a useful concept that you should understand before advancing to the next page.

This element, identified by an image of a scorpion, indicates a warning or caution. Much like in real life, when you see a scorpion, stop and look! I use the scorpion to call attention to key challenges you must address when applying algorithms.

O'Reilly Online Learning

For more than 40 years, *O'Reilly Media* has provided technology and business training, knowledge, and insight to help companies succeed.

Our unique network of experts and innovators share their knowledge and expertise through books, articles, and our online learning platform. O'Reilly's online learning platform gives you on-demand access to live training courses, in-depth learning paths, interactive coding environments, and a vast collection of text and video from O'Reilly and 200+ other publishers. For more information, visit *http://oreilly.com*.

How to Contact Us

Please address comments and questions concerning this book to the publisher:

O'Reilly Media, Inc.
1005 Gravenstein Highway North
Sebastopol, CA 95472
800-998-9938 (in the United States or Canada)
707-829-0515 (international or local)
707-829-0104 (fax)

We have a web page for this book, where we list errata, examples, and any additional information. You can access this page at *https://oreil.ly/learn-algorithms*.

Email *bookquestions@oreilly.com* to comment or ask technical questions about this book.

For news and information about our books and courses, visit *http://oreilly.com*.

Find us on Facebook: *http://facebook.com/oreilly*

Follow us on Twitter: *http://twitter.com/oreillymedia*

Watch us on YouTube: *http://youtube.com/oreillymedia*

Acknowledgments

For me, the study of algorithms is the best part of computer science. Thank you for giving me the opportunity to present this material to you. I also want to thank my wife, Jennifer, for her support on yet another book project, and my two sons, Nicholas and Alexander, who are now both old enough to learn about programming.

My O'Reilly editors—Melissa Duffield, Sarah Grey, Beth Kelly, and Virginia Wilson—improved the book by helping me organize the concepts and its explanations. My technical reviewers—Laura Helliwell, Charlie Lovering, Helen Scott, Stanley Selkow, and Aura Velarde—helped eliminate numerous inconsistencies and increase the quality of the algorithm implementations and explanations. All defects that remain are my responsibility.

## CHAPTER 1

Problem Solving

In this chapter, you will learn:

- Multiple algorithms that solve an introductory problem.
- How to consider an algorithm's performance on problem instances of size N.
- How to count the number of times a key operation is invoked when solving a given problem instance.
- How to determine order of growth as the size of a problem instance doubles.
- How to estimate *time complexity* by counting the number of key operations an algorithm executes on a problem instance of size N.
- How to estimate *space complexity* by determining the amount of memory required by an algorithm on a problem instance of size N.

Let's get started!

What Is an Algorithm?

Explaining how an algorithm works is like telling a story. Each algorithm introduces a novel concept or innovation that improves upon ordinary solutions. In this chapter I explore several solutions to a simple problem to explain the factors that affect an algorithm's performance. Along the way I introduce techniques used to analyze an algorithm's performance *independent of its implementation*, though I will always provide empirical evidence from actual implementations.

An *algorithm* is a step-by-step problem-solving method implemented as a computer program that returns a correct result in a predictable amount of time. The study of algorithms is concerned with both correctness (will this algorithm work for all input?) and performance (is this the most efficient way to solve this problem?).

Let's walk through an example of a problem-solving method to see what this looks like in practice. What if you wanted to find the largest value in an unordered list? Each Python list in Figure 1-1 is a *problem instance*, that is, the input processed by an algorithm (shown as a cylinder); the correct answer appears on the right. How is this algorithm implemented? How would it perform on different problem instances? Can you predict the time needed to find the largest value in a list of one million values?

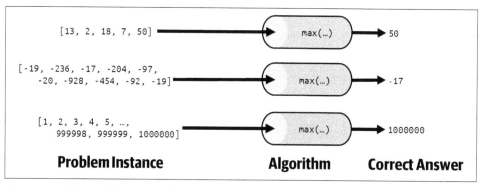

Figure 1-1. Three different problem instances processed by an algorithm

An algorithm is more than just a problem-solving method; the program also needs to complete in a predictable amount of time. The built-in Python function max() already solves this problem. Now, it can be hard to predict an algorithm's performance on problem instances containing random data, so it's worth identifying problem instances that are carefully constructed.

Table 1-1 shows the results of timing max() on two kinds of problem instances of size N: one where the list contains ascending integers and one where the list contains descending integers. While your execution may yield different results in the table, based on the configuration of your computing system, you can verify the following two statements:

- The timing for max() on ascending values is always slower than on descending values once N is *large enough*.

- As N increases ten-fold in subsequent rows, the corresponding time for max() also appears to increase ten-fold, with some deviation, as is to be expected from live performance trials.

For this problem, the maximum value is returned, and the input is unchanged. In some cases, the algorithm updates the problem instance directly instead of computing a new value—for example, sorting a list of values, as you will see in Chapter 5. In this book, N represents the size of a problem instance.

Table 1-1. Executing max() on two kinds of problem instances of size N (time in ms)

N	Ascending values	Descending values
100	0.001	0.001
1,000	0.013	0.013
10,000	0.135	0.125
100,000	1.367	1.276
1,000,000	14.278	13.419

When it comes to timing:

- You can't predict *in advance* the value of T(100,000)—that is, the time required by the algorithm to solve a problem instance of size 100,000—because computing platforms vary, and different programming languages may be used.
- However, once you empirically determine T(10,000), you can predict T(100,000)—that is, the time to solve a problem instance ten times larger—though the prediction will inevitably be inaccurate to an extent.

When designing an algorithm, the primary challenge is to ensure it is correct and works *for all input*. I will spend more time in Chapter 2 explaining how to analyze and compare the behavior of different algorithms that solve the exact same problem. The field of algorithm analysis is tied to the study of interesting, relevant problems that arise in real life. While the mathematics of algorithms can be challenging to understand, I will provide specific examples to always connect the abstract concepts with real-world problems.

The standard way to judge the efficiency of an algorithm is to count how many *computing operations* it requires. But this is exceptionally hard to do! Computers have a central processing unit (CPU) that executes *machine instructions* that perform mathematical computations (like add and multiply), assign values to CPU registers, and compare two values with each other. Modern programming languages (like C or C++) are compiled into machine instructions. Other languages (like Python or Java) are compiled into an intermediate *byte code* representation. The Python *interpreter* (which is itself a C program) executes the byte code, while built-in functions, such as min() and max(), are implemented in C and ultimately compiled into machine instructions for execution.

The Almighty Array

An *array* stores a collection of N values in a contiguous block of memory. It is one of the oldest and most dependable data structures programmers use to store multiple values. The following image represents an array of eight integers.

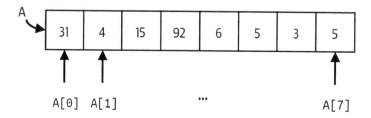

The array A has eight values indexed by their location. For example, A[0] = 31, and A[7] = 5. The values in A can be of any type, such as strings or more complicated objects.

The following are important things to know about an array:

- The first value, A[0], is at index position 0; the last is A[N-1], where N is the size of the array.
- Each array has a fixed length. Python and Java allow the programmer to determine this length at runtime, while C does not.
- One can read or update an individual location, A[i], based on the *index* position, i, which is an integer in the range from 0 to N – 1.
- An array cannot be extended (or shrunk); instead, you allocate a new array of the desired size and copy old values that should remain.

Despite their simplicity, arrays are an extremely versatile and efficient way to structure data. In Python, list objects can be considered an array, even though they are more powerful because they can grow and shrink in size over time.

It is nearly impossible to count the total number of executed machine instructions for an algorithm, not to mention that modern day CPUs can execute *billions* of instructions per second! Instead, I will count the number of times a *key operation* is invoked for each algorithm, which could be "the number of times two values in an array are compared with each other" or "how many times a function is called." In this discussion of max(), the key operation is "how many times the *less-than* (<) operator is invoked." I will expand on this counting principle in Chapter 2.

Now is a good time to lift up the hood on the max() algorithm to see why it behaves the way it does.

Finding the Largest Value in an Arbitrary List

Consider the flawed Python implementation in Listing 1-1 that attempts to find the largest value in an arbitrary list *containing at least one value* by comparing each value in A against my_max, updating my_max as needed when larger values are found.

Listing 1-1. Flawed implementation to locate largest value in list

```
def flawed(A):
    my_max = 0         ❶
    for v in A:        ❷
        if my_max < v:
            my_max = v  ❸
    return my_max
```

❶ my_max is a variable that holds the maximum value; here my_max is initialized to 0.

❷ The for loop defines a variable v that iterates over each element in A. The if statement executes once for each value, v.

❸ Update my_max if v is larger.

Central to this solution is the less-than operator (<) that compares two numbers to determine whether a value is smaller than another. In Figure 1-2, as v takes on successive values from A, you can see that my_max is updated three times to determine the largest value in A. flawed() determines the largest value in A, invoking less-than six times, once for each of its values. On a problem instance of size N, flawed() invokes less-than N times.

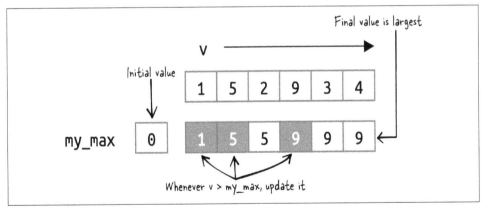

Figure 1-2. Visualizing the execution of flawed()

This implementation is flawed because it assumes that at least one value in A is greater than 0. Computing flawed([-5,-3,-11]) returns 0, which is incorrect. One common fix is to initialize my_max to the smallest possible value, such as my_max = float('-inf'). This approach is still flawed since it would return this value if A is the empty list []. Let's fix this defect now.

 The Python statement range(x,y) produces the integers from x up to, but not including, y. You can also request range(x,y,-1), which produces the integers from x counting down to, but not including, y. Thus list(range(1,7)) produces [1,2,3,4,5,6], and list(range(5,0,-1)) produces [5,4,3,2,1]. You can count by arbitrary increments, thus list(range(1,10,2)) produces [1,3,5,7,9] using a difference of 2 between values.

Counting Key Operations

Since the largest value must actually be contained in A, the correct largest() function in Listing 1-2 selects the first value of A as my_max, checking other values to see if any value is larger.

Listing 1-2. Correct function to find largest value in list

```
def largest(A):
  my_max = A[0]              ❶
  for idx in range(1, len(A)):  ❷
    if my_max < A[idx]:
      my_max = A[idx]        ❸
  return my_max
```

❶ Set my_max to the first value in A, found at index position 0.

❷ idx takes on integer values from 1 up to, but not including, len(A).

❸ Update my_max if the value in A at position idx is larger.

 If you invoke largest() or max() with an empty list, it will raise a ValueError: list index out of range exception. These runtime exceptions are programmer errors, reflecting a failure to understand that largest() requires a list with at least one value.

Now that we have a correct Python implementation of our algorithm, can you determine how many times less-than is invoked in this new algorithm? Right! N – 1 times. We have fixed the flaw in the algorithm and improved its performance (admittedly, by just a tiny bit).

Why is it important to count the uses of less-than? This is the key operation used when comparing two values. All other program statements (such as for or while loops) are arbitrary choices during implementation, based on the program language used. We will expand on this idea in the next chapter, but for now counting key operations is sufficient.

Models Can Predict Algorithm Performance

What if someone shows you a different algorithm for this same problem? How would you determine which one to use? Consider the alternate() algorithm in Listing 1-3 that repeatedly checks each value in A to see if it is larger than or equal to all other values in the same list. Will this algorithm return the correct result? How many times does it invoke less-than on a problem of size N?

Listing 1-3. A different approach to locating largest value in A

```
def alternate(A):
  for v in A:
    v_is_largest = True        ❶
    for x in A:
      if v < x:
        v_is_largest = False   ❷
        break
    if v_is_largest:
      return v                 ❸
  return None                  ❹
```

❶ When iterating over A, assume each value, v, could be the largest.

❷ If v is smaller than another value, x, stop and record that v is not greatest.

❸ If v_is_largest is true, return v since it is the maximum value in A.

❹ If A is an empty list, return None.

`alternate()` attempts to find a value, v, in A such that no other value, x, in A is greater. The implementation uses two nested `for` loops. This time it's not so simple to compute how many times less-than is invoked, because the inner `for` loop over x stops as soon as an x is found that is greater than v. Also, the outer `for` loop over v stops once the maximum value is found. Figure 1-3 visualizes executing `alternate()` on our list example.

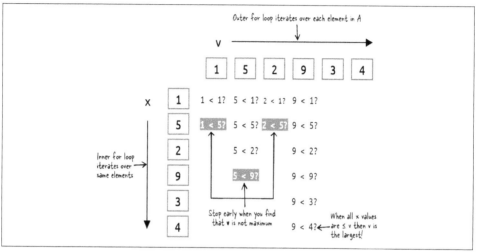

Figure 1-3. Visualizing the execution of `alternate()`

For this problem instance, less-than is invoked 14 times. But you can see that this total count depends on the specific values in the list A. What if the values were in a different order? Can you think of an arrangement of values that requires the least number of less-than invocations? Such a problem instance would be considered a *best case* for `alternate()`. For example, if the first value in A is the largest of all N values, then the total number of calls to less-than is always N. To summarize:

Best case
    A problem instance of size N that requires the least amount of work performed by an algorithm

Worst case
    A problem instance of size N that demands the most amount of work

Let's try to identify a *worst case* problem instance for `alternate()` that requires the most number of calls to less-than. More than just ensuring that the largest value is the last value in A, in a *worst case* problem instance for `alternate()`, the values in A must appear in ascending order.

Figure 1-4 visualizes a *best case* on the top where p = [9,5,2,1,3,4] and a *worst case* on the bottom where p = [1,2,3,4,5,9].

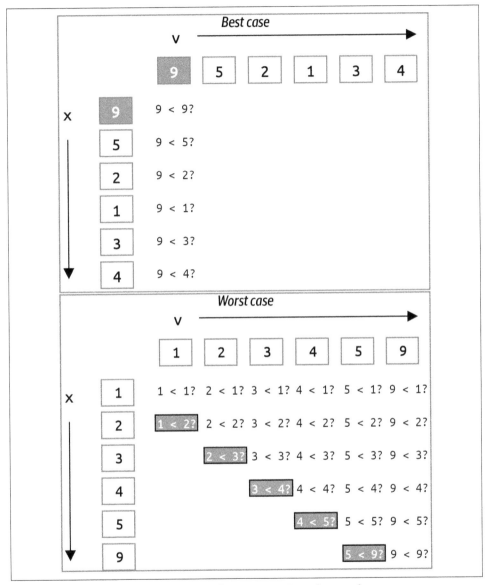

Figure 1-4. Visualizing the execution of alternate() *on best and worst cases*

In the *best case*, there are six calls to less-than; if there were N values in a *best case*, then the total number of invocations to less-than would be N. It's a bit more complicated for the *worst case*. In Figure 1-4 you can see there are a total of 26 calls to less-than when the list of N values is in ascending sorted order. With a little bit of mathematics, I can show that for N values, this count will always be $(N^2 + 3N - 2)/2$.

Table 1-2 presents empirical evidence on `largest()` and `alternate()` on *worst case* problem instances of size N.

Table 1-2. Comparing `largest()` with `alternate()` on worst case problem instances

N	Largest (# less-than)	Alternate (# less-than)	Largest (time in ms)	Alternate (time in ms)
8	7	43	0.001	0.001
16	15	151	0.001	0.003
32	31	559	0.002	0.011
64	63	2,143	0.003	0.040
128	127	8,383	0.006	0.153
256	255	33,151	0.012	0.599
512	511	131,839	0.026	2.381
1,024	1,023	525,823	0.053	9.512
2,048	2,047	2,100,223	0.108	38.161

For small problem instances, it doesn't seem bad, but as the problem instances double in size, the number of less-than invocations for `alternate()` essentially quadruples, far surpassing the count for `largest()`. The next two columns in Table 1-2 show the performance of these two implementations on 100 random trials of problem instances of size N. The completion time for `alternate()` quadruples as well.

 I measure the time required by an algorithm to process random problem instances of size N. From this set of runs, I select the quickest completion time (i.e., the smallest). This is preferable to simply averaging the total running time over all runs, which might skew the results.

Throughout this book, I am going to present tables, like Table 1-2, containing the total number of executions of a key operation (here, the less-than operator) as well as the runtime performance. Each row will represent a different problem instance size, N. Read the table from top to bottom to see how the values in each column change as the problem instance size doubles.

Counting the number of less-than invocations explains the behaviors of largest() and alternate(). As N doubles, the number of calls to less-than doubles for largest() but *quadruples* for alternate(). This behavior is consistent and you can use this information to predict the runtime performance of these two algorithms on larger-sized problems. Figure 1-5 plots the count of less-than invocations by alternate() (using the y-axis on the left) against its runtime performance (using the y-axis on the right), showing how directly they correlate with each other.

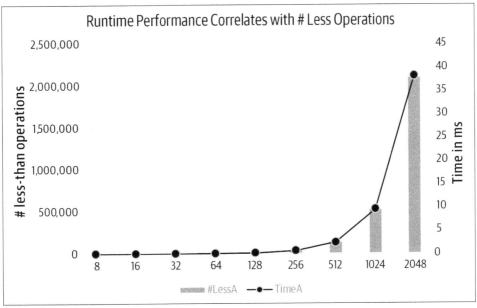

Figure 1-5. Relationship between the number of less-than operations and runtime performance

Congratulations! You've just performed a key step in algorithmic analysis: judging the relative performance of two algorithms by counting the number of times a key operation is performed. You can certainly go and implement both variations (as I did) and measure their respective runtime performance on problem instances that double in size; but it actually isn't necessary since the model predicts this behavior and confirms that largest() is the more efficient algorithm of the two.

largest() and max() are implementations of the same algorithm, but as Table 1-3 shows, largest() is always significantly slower than max(), typically four times slower. The reason is that Python is an *interpreted* language, which means it is compiled to an intermediate byte code that is executed by a Python interpreter. Built-in functions, such as max(), will always outperform Python code written for the same purpose because the built-in function is implemented within the interpreter. What you should observe is that in all cases, as N doubles, the corresponding performance of largest() and max()—for both *best case* and *worst case*—also doubles.

Table 1-3 shows it is possible to predict the time required to solve problem instances of increasing size. Once you know the runtime performance of largest() or max() on a *best* or *worst case* problem instance of size N, you can predict that the runtime performance will double as N doubles. Now let's change the problem slightly to make it more interesting.

Table 1-3. Performance of largest() and max() on best and worst cases

N	largest() worst case	max() worst case	largest() best case	max() best case
4,096	0.20	0.05	0.14	0.05
8,192	0.40	0.11	0.29	0.10
16,384	0.80	0.21	0.57	0.19
32,768	1.60	0.41	1.14	0.39
65,536	3.21	0.85	2.28	0.78
131,072	6.46	1.73	4.59	1.59
262,144	13.06	3.50	9.32	3.24
524,288	26.17	7.00	18.74	6.50

Find Two Largest Values in an Arbitrary List

Devise an algorithm that finds the *two* largest values in an arbitrary list. Perhaps you can modify the existing largest() algorithm with just a few tweaks. Why don't you take a stab at solving this modified problem and come back here with your solution? Listing 1-4 contains my solution.

Listing 1-4. Find two largest values by tweaking largest() approach

```
def largest_two(A):
  my_max,second = A[:2]              ❶
  if my_max < second:
    my_max,second = second,my_max

  for idx in range(2, len(A)):
    if my_max < A[idx]:              ❷
      my_max,second = A[idx],my_max
    elif second < A[idx]:            ❸
      second = A[idx]
  return (my_max, second)
```

❶ Ensure my_max and second are the first two values from A in descending order.

❷ If A[idx] is a newly found maximum value, then set my_max to A[idx], and second becomes the old my_max.

❸ If A[idx] is larger than second (but smaller than my_max), only update second.

largest_two() feels similar to largest(): compute my_max and second to be the first two values in A, properly ordered. Then for each of the remaining values in A (how many? N − 2, right!), if you find an A[idx] larger than my_max, adjust both my_max and second, otherwise check to see whether only second needs updating.

Counting the number of times less-than is invoked is more complicated because, once again, *it depends on the values in the problem instance.*

largest_two() performs the fewest less-than invocations when the condition of the if statement inside the for loop is true. When A contains values in ascending order, this less-than is always true, so it is invoked N − 2 times; don't forget to add 1 because of the use of less-than at the beginning of the function. In the *best case*, therefore, you only need N − 1 invocations of less-than to determine the top two values. The less-than in the elif condition is never used in the *best case*.

For largest_two(), can you construct a *worst case* problem instance? Try it yourself: it happens whenever the less-than in the if condition within the for loop is False.

I bet you can see that whenever A contains values in descending order, largest_two() requires the most invocations of less-than. In particular, for the *worst case*, less-than is used twice each time through the for loop, or 1 + 2 × (N − 2) = 2N − 3 times. Somehow this feels right, doesn't it? If you need to use less-than N − 1 times to find the largest value in A, perhaps you truly do need an additional N − 2 less-than invocations (leaving out the largest value, of course) to also find the second-largest value.

To summarize the behavior of largest_two():

- For *best case*, it finds both values with N − 1 less-than invocations.
- For *worst case*, it finds both values with 2N − 3 less-than invocations.

Are we done? Is this the "best" algorithm to solve the problem of finding the two largest values in an arbitrary list? We can choose to prefer one algorithm over another based on a number of factors:

Required extra storage
    Does an algorithm need to make a copy of the original problem instance?

Programming effort
    How few lines of code must the programmer write?

Mutable input

Does the algorithm modify the input provided by the problem instance *in place*, or does it leave it alone?

Speed

Does an algorithm outperform the competition, independent of the provided input?

Let's investigate three different algorithms to solve this exact same problem, shown in Listing 1-5. `sorting_two()` creates a new list containing the values in A in descending order, grabs the first two values, and returns them as a tuple. `double_two()` uses `max()` to find the maximum value in A, removes it from a copy of A, and then uses `max()` of that reduced list to find the second largest. `mutable_two()` finds the location of the largest value in A and removes it from A; then it sets `second` to the largest value remaining in A before reinserting the `my_max` value into its original location. The first two algorithms require extra storage, while the third modifies its input: all three only work on problem instances containing more than one value.

Listing 1-5. Three different approaches using Python utilities

```
def sorting_two(A):
  return tuple(sorted(A, reverse=True)[:2])      ❶

def double_two(A):
  my_max = max(A)                                ❷
  copy = list(A)
  copy.remove(my_max)                            ❸
  return (my_max, max(copy))                     ❹

def mutable_two(A):
  idx = max(range(len(A)), key=A.__getitem__)    ❺
  my_max = A[idx]                                ❻
  del A[idx]
  second = max(A)                                ❼
  A.insert(idx, my_max)                          ❽
  return (my_max, second)
```

❶ Create a new list by sorting A in descending order and return its top two values.

❷ Use built-in `max()` function to find largest.

❸ Create a copy of the original A, and remove `my_max`.

❹ Return a tuple containing `my_max` and the largest value in `copy`.

❺ This Python trick finds the *index location* of the maximum value in A, rather than the value itself.

❻ Record `my_max` value and delete it from A.

❼ Now find `max()` of remaining values.

❽ Restore A by inserting `my_max` to its original location.

These different approaches do not directly use less-than because they rely on existing Python libraries. Both `sorting_two()` and `double_two()` make a copy of the array, A, which seems unnecessary, since `largest_two()` doesn't do this. In addition, it seems excessive to *sort the entire list* just to find the top two largest values. In the same way that I count operations when analyzing runtime performance, I will evaluate the *extra storage* used by an algorithm—for both of these approaches, the amount of storage is directly proportional to N. The third approach, `mutable_two()`, briefly updates A by deleting its maximum value, only to add it back later. The caller might be surprised to see that the original list was modified.

With a bit of Python expertise, I can compute exactly how many times less-than is invoked using a special `RecordedItem` class.[1] Table 1-4 shows that `double_two()` invokes the most less-than operations when the values are in ascending order, while `largest_two()` (and others) perform the most less-than operations when the values are in descending order. In the last column, labeled "Alternating," the 524,288 values are arranged with even numbers in ascending order, alternating with odd numbers in descending order: for N = 8, the input would be [0,7,2,5,4,3,6,1].

Table 1-4. Performance of different approaches on 524,288 ascending and descending values

Algorithm	Ascending	Descending	Alternating
largest_two	524,287	1,048,573	1,048,573
sorting_two	524,287	524,287	2,948,953
double_two	1,572,860	1,048,573	1,048,573
mutable_two	1,048,573	1,048,573	1,048,573
tournament_two	524,305	524,305	524,305

The `tournament_two()` algorithm I describe next has the fewest number of less-than invocations regardless of input. Basketball fans will find its logic familiar.

---

1 The `RecordedItem` wrapper class overrides the `__lt__()` less-than operator to count whenever it (or the `__gt__()` *greater-than* operator) is invoked.

If you determine the *worst case* problem instance for an algorithm that solves a given problem, perhaps a different algorithm solving the same problem would not be so negatively affected by that problem instance. Different algorithms can have different weaknesses that you can uncover with diligent analysis.

Tournament Algorithm

A single-elimination tournament consists of a number of teams competing to be the champion. Ideally, the number of teams is a power of 2, like 16 or 64. The tournament is built from a sequence of rounds where all remaining teams in the tournament are paired up to play a match; match losers are eliminated, while winners advance to the next round. The final team remaining is the tournament champion.

Consider the problem instance p = [3,1,4,1,5,9,2,6] with N = 8 values. Figure 1-6 shows the single-elimination tournament whose initial round compares eight neighboring values using less-than; larger values advance in the tournament.[2] In the Elite Eight round, four values are eliminated, leaving values [3,4,9,6]. In the Final Four round, values [4,9] advance, and eventually 9 is declared the champion.

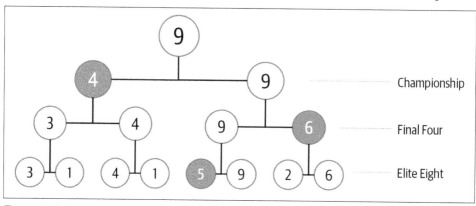

Figure 1-6. A tournament with eight initial values

In this tournament, there are seven less-than invocations (i.e., one for each match), which should be reassuring, since this means the largest value is found with N – 1 uses of less-than, as we had discussed earlier. If you store each of these N – 1 matches, then you can quickly locate the second-largest value, as I now show.

---

2 If a match contains two equal values, then only one of these values advances.

Where can the second-largest value be "hiding" once 9 is declared the champion? Start with 4 as the candidate second-largest value, since this was the value that lost in the Championship round. But the largest value, 9, had two earlier matches, so you must check the other two losing values—value 6 in the Final Four round and value 5 in the Elite Eight round. Thus the second-largest value is 6.

For eight initial values, you need just 2 additional less-than invocations—(is 4 < 6?) and (is 6 < 5?)—to determine that 6 is the second-largest value. It's no coincidence that $8 = 2^3$ and you need $3 - 1 = 2$ comparisons. It turns out that for $N = 2^K$, you need an additional $K - 1$ comparisons, where K is the number of rounds.

When there are $8 = 2^3$ initial values, the algorithm constructs a tournament with 3 rounds. Figure 1-7 visualizes a five-round tournament consisting of 32 values. To double the number of values in the tournament, you only need one additional round of matches; in other words, with each new round K, you can add 2^K more values. Want to find the largest of 64 values? You only need 6 rounds since $2^6 = 64$.

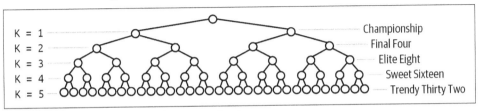

Figure 1-7. A tournament with 32 initial values

To determine the number of rounds for any N, turn to the *logarithm* function, $\log()$, which is the opposite of the exponent function. With $N = 8$ initial values, there are 3 rounds required for the tournament, since $2^3 = 8$ and $\log_2(8) = 3$. In this book—and traditionally in computer science—the $\log()$ operator is in base 2.

Most handheld calculators compute $\log()$ in base 10. The function $\ln()$ represents the natural logarithm in base e (which is approximately 2.718). To quickly compute $\log(N)$ in base 2 using a calculator (or in Microsoft Excel), compute $\log(N)/\log(2)$.

When N is a power of 2—like 64 or 65,536—the number of rounds in a tournament is $\log(N)$, which means the number of additional less-than invocations is $\log(N) - 1$. The algorithm implemented in Listing 1-6 minimizes the invocations of less-than by using extra storage to record the results of all matches.

Listing 1-6. Algorithm to find two largest values in A using tournament

```
def tournament_two(A):
  N = len(A)
  winner = [None] * (N-1)          ❶
  loser = [None] * (N-1)
  prior = [-1] * (N-1)             ❷

  idx = 0
  for i in range(0, N, 2):         ❸
    if A[i] < A[i+1]:
      winner[idx] = A[i+1]
      loser[idx] = A[i]
    else:
      winner[idx] = A[i]
      loser[idx] = A[i+1]
    idx += 1

  m = 0                            ❹
  while idx < N-1:
    if winner[m] < winner[m+1]:    ❺
      winner[idx] = winner[m+1]
      loser[idx]  = winner[m]
      prior[idx]  = m+1
    else:
      winner[idx] = winner[m]
      loser[idx]  = winner[m+1]
      prior[idx]  = m
    m += 2                         ❻
    idx += 1

  largest = winner[m]
  second = loser[m]                ❼
  m = prior[m]
  while m >= 0:
    if second < loser[m]:          ❽
      second = loser[m]
    m = prior[m]

  return (largest, second)
```

❶ These arrays store the winners and losers of match idx; there will be N − 1 of them in the tournament.

❷ When a value advances in match m, prior[m] records earlier match, or −1 when it was initial match.

❸ Initialize the first N/2 winner/loser pairs using N/2 invocations of less-than. These represent the matches in the lowest round.

❹ Pair up winners to find a new winner, and record `prior` match index.

❺ An additional $N/2 - 1$ invocations of less-than are needed.

❻ Advance `m` by 2 to find next pair of winners. When `idx` reaches $N - 1$, `winner[m]` is largest.

❼ Initial candidate for second largest, but must check all others that lost to `largest` to find actual second largest.

❽ No more than $\log(N) - 1$ additional invocations of less-than.

Figure 1-8 shows the execution of this algorithm. After the initialization step, the N values from the original array, `A`, are separated into N/2 `winners` and `losers`; in the example from Figure 1-6, there are four pairs. During each advance step in the `while` loop, the winner/loser of two consecutive matches, `m` and `m+1`, are placed in `winner[idx]` and `loser[idx]` respectively; `prior[idx]` records the prior match from which the winner came (as drawn by an arrow from right to left). After three steps, all match information is stored, and then the algorithm inspects the losers of all prior matches for the winner. You can visualize this by following the arrows backward until they stop. You can see that the candidate second-largest value is found in `loser[6]`: with just two less-than invocations with `loser[5]` and `loser[2]`, it determines which one is largest.

I have just sketched an algorithm to compute the largest and second-largest value in `A` using just $N - 1 + \log(N) - 1 = N + \log(N) - 2$ less-than invocations for any N that is a power of 2. Is `tournament_two()` practical? Will it outperform `largest_two()`? If you only count the number of times less-than is invoked, `tournament_two()` should be faster. `largest_two()` requires 131,069 less-than operations on problems of size $N = 65,536$, while `tournament_two()` only requires $65,536 + 16 - 2 = 65,550$, just about half. But there is more to this story.

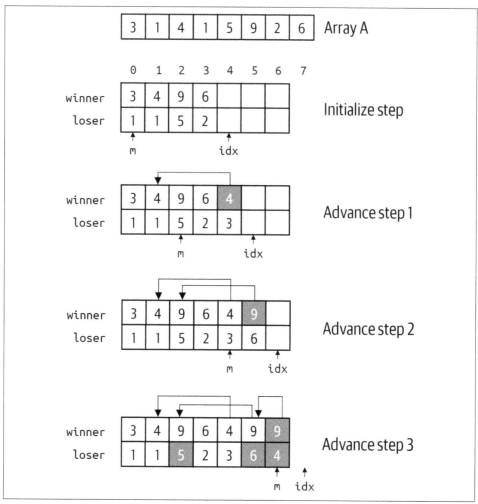

Figure 1-8. Step-by-step execution of tournament algorithm

Table 1-5 reveals that `tournament_two()` is significantly slower than any of its competitors! Let's record the total time it takes to solve 100 random problem instances (of size N growing from 1,024 to 2,097,152). While I'm at it, I will include the performance results of the different algorithms from Listing 1-5. Note that if you run the sample code on your computer, your individual results will be different, but *the overall trend* in each column will remain the same.

Table 1-5. Comparing runtime performance (in ms) of all four algorithms

N	double_two	mutable_two	largest_two	sorting_two	tournament_two
1,024	0.00	0.01	0.01	0.01	0.03
2,048	0.01	0.01	0.01	0.02	0.05
4,096	0.01	0.02	0.03	0.03	0.10
8,192	0.03	0.05	0.05	0.08	0.21
16,384	0.06	0.09	0.11	0.18	0.43
32,768	0.12	0.20	0.22	0.40	0.90
65,536	0.30	0.39	0.44	0.89	1.79
131,072	0.55	0.81	0.91	1.94	3.59
262,144	1.42	1.76	1.93	4.36	7.51
524,288	6.79	6.29	5.82	11.44	18.49
1,048,576	16.82	16.69	14.43	29.45	42.55
2,097,152	35.96	38.10	31.71	66.14	...

Table 1-5 can be overwhelming to look at, since it just looks like a wall of numbers. If you run these functions on a different computer—perhaps with less memory or a slower CPU—your results might be quite different; however, there are some trends that should reveal themselves no matter on what computer you execute. For the most part, as you read down any column, the time to execute more or less doubles as the problem size doubles.

There are some unexpected situations in this table: note that double_two() starts out being the fastest of the five solutions, but eventually (once N > 262,144), larg est_two() becomes the fastest to complete. The clever tournament_two() approach is by far the slowest, simply because it needs to allocate ever-larger storage arrays to be processed. It is so slow, I do not even run it on the largest problem instance because it will take so long.

To make sense of these numbers, Figure 1-9 visualizes the runtime trends as the problem size instance grows ever larger.

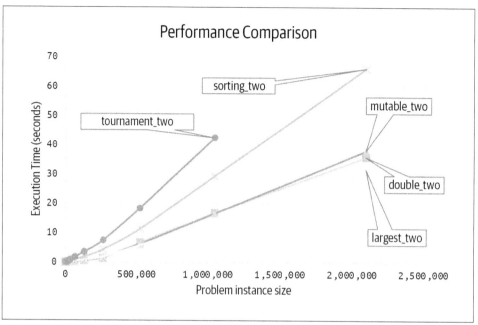

Figure 1-9. Runtime performance comparison

This image reveals more details about the runtime performance of these five approaches:

- You can see that the performances of `mutable_two()`, `double_two()`, and `larg est_two()` are all more similar to each other than the other two approaches. It's almost like they are all in the same "family," all on a straight-line trajectory that appears quite predictable.

- `tournament_two()` is the least efficient, and it noticeably behaves differently from the other approaches. Given that there are so few data points, it is not clear whether it is "curving upward" to greater inefficiencies or whether it also will follow a straight line.

- `sorting_two()` appears to do better than `tournament_two()` but is slower than the other three approaches. Will it curve further upward, or eventually straighten out?

To understand why these lines are shaped the way they are, you need to learn the two fundamental concepts that explain the inherent *complexity* of an algorithm.

Time Complexity and Space Complexity

It can be hard to count the number of elementary operations (such as addition, variable assignment, or control logic), because of the difference in programming languages, plus the fact that some languages, such as Java and Python, are executed by an interpreter. But if you *could* count the total number of elementary operations executed by an algorithm, then you would see that the total number of operations varies based on the size of the problem instance. The goal of *time complexity* is to come up with a function C(N) that counts the number of elementary operations performed by an algorithm as a function of N, the size of a problem instance.

Assuming that each elementary operation takes a fixed amount of time, t, based on the CPU executing the operation, I can model the time to perform the algorithm as T(N) = t × C(N). Listing 1-7 confirms the insight that the *structure* of a program is critical. For functions f0, f1, f2, and f3, you can exactly compute how many times each one executes the operation ct = ct + 1 based on the input size, N. Table 1-6 contains the counts for a few values of N.

Listing 1-7. Four different functions with different performance profiles

```
def f0(N):        def f1(N):            def f2(N):            def f3(N):
  ct = 0            ct = 0                ct = 0                ct = 0
  ct = ct + 1       for i in range(N):    for i in range(N):    for i in range(N):
  ct = ct + 1         ct = ct + 1           ct = ct + 1           for j in range(N):
  return ct         return ct             ct = ct + 1             ct = ct + 1
                                          ct = ct + 1           return ct
                                          ct = ct + 1
                                          ct = ct + 1
                                          ct = ct + 1
                                          ct = ct + 1
                                        return ct
```

The count for f0 is always the same, independent of N. The count for f2 is always seven times greater than f1, and both of them double in size as N doubles. In contrast, the count for f3 increases far more rapidly; as you have seen before, as N doubles, the count for f3(N) quadruples. Here, f1 and f2 are more similar to each other than they are to f3. In the next chapter, we will explain the importance of for loops and nested for loops when evaluating an algorithm's performance.

Table 1-6. Counting operations in four different functions

N	f0	f1	f2	f3
512	2	512	3,584	262,144
1,024	2	1,024	7,168	1,048,576
2,048	2	2,048	14,336	4,194,304

When evaluating an algorithm, we also have to consider *space complexity*, which accounts for extra memory required by an algorithm based on the size, N, of a problem instance. *Memory* is a generic term for data stored in the file system or the RAM of a computer. largest_two() has minimal space requirements: it uses two variables, my_max and second, and an iterator variable, idx. No matter the size of the problem instance, its extra space never changes. This means the space complexity is *independent of the size of the problem instance*, or constant; mutable_two() has similar behavior. In contrast, tournament_two() allocated three arrays—winner, loser, and prior —all of size N – 1. As N increases, the total extra storage increases in a manner that is *directly proportional to the size of the problem instance*.[3] Building the tournament structure is going to slow tournament_two() down, when compared against larg est_two(). Both double_two() and sorting_two() make a copy of the input, A, which means their storage usage is much more like tournament_two() than larg est_two(). Throughout this book, I will evaluate both the time complexity and space complexity of each algorithm.

If you review Table 1-5, you can see that the timing results for the column largest_two more or less double in subsequent rows; columns double_two and mutable_two behave similarly, as I have already observed. This means that the total time appears to be *directly proportional to the size of the problem instance*, which is doubling in subsequent rows. This is an important observation, since these functions are more efficient than sorting_two(), which appears to follow a different, less-efficient trajectory. tournament_two() is still the least efficient, with a runtime performance that more than doubles, growing so rapidly that I don't bother executing it for large problem instances.

As a computer scientist, I cannot just proclaim that the performance curves of largest_two() and mutable_two() "look the same." I need to rely on a formal theory and notation to capture this idea. In the next chapter, I will present the mathematical tools necessary to analyze algorithm behavior properly.

Summary

This chapter provided an introduction to the rich and diverse field of algorithms. I showed how to model the performance of an algorithm on a problem instance of size N by counting the number of key operations it performs. You can also empirically evaluate the runtime performance of the implementation of an algorithm. In both cases, you can determine the order of growth of the algorithm as the problem instance size N doubles.

---

3 That is, storage in addition to the data encoding the problem instance, which is not counted as part of the space complexity of any algorithm.

I introduced several key concepts, including:

- *Time complexity* as estimated by counting the number of key operations executed by an algorithm on a problem instance of size N.
- *Space complexity* as estimated by the amount of memory required when an algorithm executes on a problem instance of size N.

In the next chapter, I will introduce the mathematical tools of *asymptotic analysis* that will complete my explanation of the techniques needed to properly analyze algorithms.

Challenge Exercises

1. *Palindrome word detector*: A palindrome word reads the same backward as forward, such as *madam*. Devise an algorithm that validates whether a word of N characters is a palindrome. Confirm empirically that it outperforms the two alternatives in Listing 1-8:

   Listing 1-8. Four different functions with different performance profiles

   ```
   def is_palindrome1(w):
       """Create slice with negative step and confirm equality with w."""
       return w[::-1] == w

   def is_palindrome2(w):
       """Strip outermost characters if same, return false when mismatch."""
       while len(w) > 1:
           if w[0] != w[-1]:        # if mismatch, return False
               return False
           w = w[1:-1]              # strip characters on either end; repeat

       return True                  # must have been palindrome
   ```

   Once you have this problem working, modify it to detect palindrome strings with spaces, punctuation, and mixed capitalization. For example, the following string should classify as a palindrome: "A man, a plan, a canal. Panama!"

2. *Linear time median*: A wonderful algorithm exists that efficiently locates the median value in an arbitrary list (for simplicity, assume size of list is odd). Review the code in Listing 1-9 and count the number of times less-than is invoked, using RecordedItem values as shown in the chapter. This implementation rearranges the arbitrary list as it processes.

Listing 1-9. A linear-time algorithm to compute the median value in an unordered list

```python
def partition(A, lo, hi, idx):
  """Partition using A[idx] as value."""
  if lo == hi: return lo

  A[idx],A[lo] = A[lo],A[idx]     # swap into position
  i = lo
  j = hi + 1
  while True:
    while True:
      i += 1
      if i == hi: break
      if A[lo] < A[i]: break

    while True:
      j -= 1
      if j == lo: break
      if A[j] < A[lo]: break

    if i >= j: break
    A[i],A[j] = A[j],A[i]

  A[lo],A[j] = A[j],A[lo]
  return j

def linear_median(A):
  """
  Efficient implementation that returns median value in arbitrary list,
  assuming A has an odd number of values. Note this algorithm will
  rearrange values in A.
  """
  lo = 0
  hi = len(A) - 1
  mid = hi // 2
  while lo < hi:
    idx = random.randint(lo, hi)     # select valid index randomly
    j = partition(A, lo, hi, idx)

    if j == mid:
      return A[j]
    if j < mid:
      lo = j+1
    else:
      hi = j-1
  return A[lo]
```

Implement a different approach (which requires extra storage) that creates a sorted list from the input and selects the middle value. Compare its runtime performance with `linear_median()` by generating a table of runtime performance.

3. *Counting Sort*: If you know that an arbitrary list, A, only contains nonnegative integers from 0 to M, then the following algorithm will properly sort A using just an extra storage of size M.

Listing 1-10 has nested loops—a `for` loop within a `while` loop. However, you can demonstrate that `A[pos+idx]` = v only executes N times.

Listing 1-10. A linear-time Counting Sort algorithm

```
def counting_sort(A, M):
  counts = [0] * M
  for v in A:
    counts[v] += 1

  pos = 0
  v = 0
  while pos < len(A):
    for idx in range(counts[v]):
      A[pos+idx] = v
    pos += counts[v]
    v += 1
```

Conduct a performance analysis to demonstrate that the time to sort N integers in the range from 0 to M doubles as the size of N doubles.

You can eliminate the inner `for` loop, and improve the performance of this operation, using the ability in Python to replace a sublist using `sublist[left:right]` = `[2,3,4]`. Make the change and empirically validate that it, too, doubles as N doubles, while yielding a 30% improvement in speed.

4. Modify tournament algorithm to work with an odd number of values.

5. Will the code in Listing 1-11 correctly locate the two largest values in A?

Listing 1-11. Another attempt to try to compute two largest values in unordered list

```
def two_largest_attempt(A):
  m1 = max(A[:len(A)//2])
  m2 = max(A[len(A)//2:])
  if m1 < m2:
    return (m2, m1)
  return (m1, m2)
```

Explain the circumstances when this code works correctly and when it fails.

CHAPTER 2

Analyzing Algorithms

In this chapter, you will learn:

- How to use the *Big O* notation to classify the performance of algorithms (in time or storage)
- Several *performance classes*, including:
  - O(1) or *constant*
  - O(log N) or *logarithmic*
  - O(N) or *linear*
  - O(N log N)[1]
  - O(N^2) or *quadratic*
- How *asymptotic analysis* estimates in terms of N the time (or storage space) required by an algorithm to process a problem instance of size N.
- How to work with arrays whose values appear in ascending, sorted order.
- The *Binary Array Search* algorithm to locate values in a sorted array.

This chapter introduces the terminology and notation used by theoreticians and practitioners alike in modeling the performance of algorithms in terms of computational performance and resource usage. When evaluating the runtime performance of your software program, you might be perfectly satisfied, in which case you can continue to use the application as is. But if you want to improve runtime performance,

---

1 Pronounced using three syllables, like en-log-en.

this book shows you where to start—with the program's data structures and algorithms. You are faced with some specific questions:

Am I solving a specific problem in the most efficient way?
There may be other algorithms that would significantly improve performance.

Am I implementing an algorithm in the most efficient way?
There can be hidden performance costs that can be eliminated.

Should I just buy a faster computer?
The exact same program will have different runtime performance based on the computer on which it runs. In this chapter, I explain how computer scientists have developed analysis techniques that account for regular hardware improvements.

I start by showing how to model the runtime performance of a program on everincreasing problem instance sizes. The runtime performance of an algorithm on small problem instance sizes can be hard to measure accurately because it could be sensitive to the actual values in the problem instance or to the resolution of the computer timers. Once your program processes a *sufficiently large problem instance*, you can develop models to classify its runtime behavior using empirical models.

Using Empirical Models to Predict Performance

I'd like to start with an example that shows how theoretical analysis is decidedly practical in real software systems. Imagine you are in charge of designing an application as part of a nightly batch job to process a large data set each night; the task is launched at midnight and must complete before 6:00 a.m. The data set contains several million values and is expected to double in size over the next five years.

You have built a working prototype but have only tested it on multiple small data sets, with 100, 1,000, and 10,000 values each. Table 2-1 presents the runtime performance of your prototype on these data sets.

Table 2-1. Prototype runtime performance

N	Time (seconds)
100	0.063
1,000	0.565
10,000	5.946

Can these preliminary results predict the performance of your prototype on larger problem instances, such as 100,000 or even 1,000,000? Let's build mathematical models *from this data alone* to define a function $T(N)$ that predicts the runtime performance for a given problem instance size. An accurate model will compute a $T(N)$ that is close to the three values in Table 2-1, as well as predict higher values of N, as shown in Table 2-2 (which repeats these three time results in brackets).

You may have used a software tool, such as Maple (*https://maplesoft.com*) or Microsoft Excel (*https://microsoft.com/excel*), to compute a *trendline* (also called a line of best fit) for sample data. The popular SciPy library for mathematics, science, and engineering can develop these trendline models. Listing 2-1 uses `scipy` to try to find a linear model, TL(N) = a × N + b, where a and b are constants. `curve_fit()` will return the (a, b) coefficients to use with the linear model based on the available empirical data encoded in lists `xs` and `ys`.

Listing 2-1. Calculate models based on partial data

```
import numpy as np
from scipy.optimize import curve_fit

def linear_model(n, a, b):
  return a*n + b

# Sample data
xs = [100, 1000, 10000]
ys = [0.063, 0.565, 5.946]

# Coefficients are returned as first argument
[(a,b), _]  = curve_fit(linear_model, np.array(xs), np.array(ys))
print('Linear = {}*N + {}'.format(a, b))
```

The resulting model is the formula TL(N) = 0.000596 × N − 0.012833. As you can see from Table 2-2, this model is inaccurate because as the problem size increases, it significantly underestimates the actual runtime performance of the prototype. Another possible model is a *quadratic polynomial*, where N is raised to the power of 2:

```
    def quadratic_model(n, a, b):
      return a*n*n + b*n;
```

With `quadratic_model`, the goal is to find TQ(N) = a × N^2 + b × N, where a and b are constants. Using the approach in Listing 2-1, the formula is TQ(N) = 0.000000003206 × N^2 + 0.000563 × N. Table 2-2 shows that as the problem size increases, this model significantly overestimates the actual runtime performance, so it is also not accurate.

Many of these constants are quite small, like 0.000000003206, which is 3.206×10^{-9}. The reason is that the problems solved by algorithms involve problem instances where N = 1,000,000 or higher. Note that $(1,000,000)^2 = 10^{12}$, so be prepared to see both very small and very large constants.

The final column in Table 2-2 contains the predicted result using a third mathematical model, $TN(N) = a \times N \times \log(N)$, which uses the logarithm function (in base 2) and in which a is a constant. The result is $TN(N) = 0.0000448 \times N \times \log(N)$. For N = 10,000,000, the estimate computed by $TN(N)$ is within 5% of the actual value.

Table 2-2. Comparing different mathematical models against actual performance

N	Time (seconds)	TL	TQ	TN
100	[0.063]	0.047	0.056	0.030
1,000	[0.565]	0.583	0.565	0.447
10,000	[5.946]	5.944	5.946	5.955
100,000	65.391	59.559	88.321	74.438
1,000,000	860.851	595.708	3769.277	893.257
10,000,000	9879.44	5957.194	326299.837	10421.327

The linear model, TL, underestimates the total time, while the quadratic model, TQ, overestimates it. For N = 10,000,000, TL declares it will take 5,957 seconds (about 100 minutes), but TQ declares it will take 326,300 seconds (about 91 hours). TN does a better job in predicting the performance, estimating 10,421 seconds (about 2.9 hours) against the actual performance of 9,879 seconds (2.75 hours).

The prototype completes its processing overnight—that's a relief!—but you must review the code for your prototype to see the algorithms and data structures it employs, so you can guarantee this result regardless of the problem instance being solved.

Why does the formula $a \times N \times \log(N)$ model the behavior so well? It has to do with the fundamental algorithms used within your prototype application. These three kinds of models—linear, quadratic, and N log N—appear regularly when analyzing algorithms. Let's try one more example to demonstrate a surprising result discovered about fifty years ago.[2]

Multiplication Can Be Faster

Consider two examples, shown in Listing 2-2, for multiplying two N-digit integers using an algorithm most of us learned in grade school. While I do not precisely define this algorithm, you see that it creates N products, listed below the original numbers, which are totaled to compute the final answer.

---

2 A fast multiplication algorithm was discovered in 1960 by a 23-year old student at Moscow State University named Anatoly Karatsuba. Python uses this algorithm when multiplying very large integers.

Listing 2-2. Using grade-school algorithm to multiply two N-digit integers

```
    456                 123456
  x 712               x 712835
  - - -               - - - - - -
    912                 617280
    456                 370368
  3192                  987648
  - - - - - -           246912
324672                  123456
                        864192

                      - - - - - - - - - - -
                      88003757760
```

When multiplying two 3-digit integers, you need 9 single-digit multiplications. For 6-digit integers, you need 36 single-digit multiplications. Using this algorithm with two N-digit integers requires N^2 single-digit multiplications. Another observation is that *when you double the number of digits in the integers being multiplied, you need four times as many single-digit multiplications.* I'm not even counting all the other work to do (like additions) because single-digit multiplication is the *key operation.*

A computer's CPU provides efficient operations to multiply fixed-size 32-bit or 64-bit integers, but it has no ability to deal with larger integers. Python automatically upgrades large integer values to a *Bignum* structure, which allows integers to grow to any size necessary. This means you can measure the runtime performance when multiplying two N-digit numbers. Table 2-3 derives three models based on the first five rows of timing results of multiplying two N-digit integers (shown in brackets).

Table 2-3. Multiplying two N-digit integers

N	Time (seconds)	TL	TQ	Karatsuba	TKN
256	[0.0009]	-0.0045	0.0017	0.0010	0.0009
512	[0.0027]	0.0012	0.0038	0.0031	0.0029
1,024	[0.0089]	0.0126	0.0096	0.0094	0.0091
2,048	[0.0280]	0.0353	0.0269	0.0282	0.0278
4,096	[0.0848]	0.0807	0.0850	0.0846	0.0848
8,192	0.2524	0.1716	0.2946	0.2539	0.2571
16,384	0.7504	0.3534	1.0879	0.7617	0.7765
32,768	2.2769	0.7170	4.1705	2.2851	2.3402
65,536	6.7919	1.4442	16.3196	6.8554	7.0418
131,072	20.5617	2.8985	64.5533	20.5663	21.1679
262,144	61.7674	5.8071	256.7635	61.6990	63.5884

TL is a linear model, while TQ is a quadratic model. Karatsuba is the unusual formula $a \times N^{1.585}$, and the improved model $TKN(N) = a \times N^{1.585} + b \times N$, where a and b are constants.[3] TL significantly underestimates the time. TQ significantly overestimates the time, which is surprising since earlier intuition suggests that when N doubles, the time to perform should increase fourfold, an essential characteristic of quadratic models. These other models more accurately predict the performance of multiplying N-digit integers in Python, which uses an advanced, more efficient Karatsuba multiplication algorithm for large integers.

The approach used to generate Table 2-2 and Table 2-3 is a good start, but it is limited since it is indirect, based only on runtime performance and not by reviewing the code. Throughout this book, I will describe the implementations of the algorithms, and based on the structure of the code, I can identify the appropriate formula to model the performance of an algorithm.

Performance Classes

When different algorithms solve the exact same problem, it is sometimes possible to identify which one will be the most efficient simply by classifying its performance using mathematical models. Often algorithms are described using phrases like "Complexity is $O(N^2)$" or "Worst-case performance is $O(N \log N)$." To explain this terminology, I want to start with the image in Figure 2-1, which might be a familiar one if you have read a book, or an online resource, that discusses algorithm analysis.

The goal is to find a model that predicts *the worst runtime performance* for a given problem instance, N. In mathematics, this is known as an *upper bound*—think of the phrase "the algorithm will never work harder than this." A corresponding concept, the *lower bound*, represents the minimum runtime performance—in other words, "the algorithm must always work at least this much."

To explain the concepts of lower and upper bounds, consider how a car's speedometer is a model that computes an indicated speed as an approximation of the car's true speed. The indicated speed must *never be less than the true speed* so the driver can abide by speed limits. This represents the mathematical concept of a *lower bound*. On the high end, the indicated speed is allowed to be up to 110% of the true speed plus 4 kilometers per hour.[4] This represents the mathematical concept of an *upper bound*.

The true speed for the car must always be larger than the lower bound and smaller than the upper bound.

---

3 The exponent, 1.585, is the approximate value of $\log(3)$ in base 2, which is 1.58496250072116.

4 This is a European Union requirement; in the United States, it is sufficient for the speedometer to be accurate to within plus or minus 5 miles per hour.

In Figure 2-1, the three curves—TL, TQ, and TKN—represent the model predictions, while black squares represent the individual actual performance when multiplying two N-digit integers. While TQ(N) is an upper bound on the actual performance (since TQ(N) > Time for all values of N), it is highly inaccurate, as you can see from Figure 2-1.

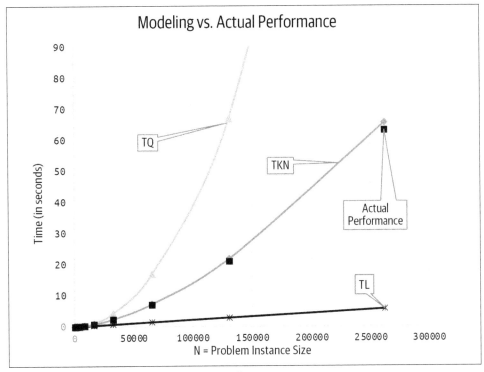

Figure 2-1. Comparing models against performance

If you go back and review Table 2-3, observe that for all N values greater than or equal to the threshold problem instance size of 8,192, TKN(N) is greater than the actual performance reported for N, while remaining much closer to the actual values. This evidence is a clear indication that often the real behavior will stabilize once N is "large enough," which will depend on each algorithm and the way it is implemented.

It might seem that TL(N) models the lower bound on the actual performance, since TL(N) < Time for all values of N. However, as N increases, it becomes further and further away from the runtime performance, essentially rendering it useless as a model of the algorithm's runtime performance. The Karatsuba formula a × N$^{1.585}$, whose values appear in Table 2-3, provide a more accurate lower bound.

If you run these programs on different computers, the numeric details shown in Table 2-3 will change—the runtime performance can be slower or faster; the a and b coefficients of TK() will change; the threshold problem instance size above which TK(N) stabilizes could be lower or higher. What would remain unchanged is the exponent 1.585 in the model, since the structure of the Karatsuba fast multiplication algorithm determines how it will perform. No supercomputer will somehow make the Karatsuba implementation suddenly behave in a way modeled by the linear TL(N) model.

We can now tackle *asymptotic analysis*, an approach that lets us eliminate any knowledge about the actual computer when evaluating the performance of an algorithm. Powerful computers can make code run faster, but they cannot bend the laws of asymptotic analysis.

Asymptotic Analysis

The concept of an *additive constant* is common in many real-world scenarios, like the speedometer I just discussed. It's what we mean when we say, "I'll be there in 40 minutes, give or take 5 minutes."

Asymptotic analysis takes this idea further and introduces the notion of a *multiplicative constant* to analyze algorithms. If you've heard of *Moore's Law*, you will find this concept familiar. Gordon Moore, the CEO and cofounder of Intel corporation, predicted in 1965 that the number of components per integrated circuit would double every year for a decade; in 1975 he revised this prediction to doubling every two years. This prediction was valid for over 40 years and explains why the speed of computers essentially doubles every two years. A multiplicative constant applied to computing means you can find an older computer where *the same program runs one thousand times slower* (or even worse) than on a modern computer.

Consider two algorithms that solve the same problem. Using the techniques I have already shown, assume that algorithm X requires 5N operations on problems of size N, while algorithm Y requires $2020 \times \log(N)$ operations to solve the same problem. Is algorithm X more efficient than Y?

You have two computers on which you execute implementations of these algorithms: computer C_{fast} is two times faster than C_{slow}. Figure 2-2 reports the number of operations for each algorithm on a problem instance of size N. It also shows performance of X and Y on C_{fast} (i.e., columns labeled X_{fast} and Y_{fast}) and the performance of X on C_{slow} (i.e., column labeled X_{slow}).

	# Operations		Time to Execute			
N	X	Y	X_{slow}	X_{fast}	Y_{fast}	$X_{fastest}$
4	20	4,040	0.0	0.0	2.7	0.0
8	40	6,060	0.0	0.0	4.0	0.0
16	80	8,080	0.1	0.0	5.4	0.0
32	160	10,100	0.1	0.1	6.7	0.0
64	320	12,120	0.2	0.1	8.1	0.0
128	640	14,140	0.4	0.2	9.4	0.0
256	1,280	16,160	0.9	0.4	10.8	0.0
512	2,560	18,180	1.7	0.9	12.1	0.0
1,024	5,120	20,200	3.4	1.7	13.5	0.0
2,048	10,240	22,220	6.8	3.4	14.8	0.0
4,096	20,480	24,240	13.7	6.8	16.2	0.0
8,192	40,960	26,260	27.3	13.7	17.5	0.1
16,384	81,920	28,280	54.6	27.3	18.9	0.1
32,768	163,840	30,300	109.2	54.6	20.2	0.2
65,536	327,680	32,320	218.5	109.2	21.5	0.4
131,072	655,360	34,340	436.9	218.5	22.9	0.9
262,144	1,310,720	36,360	873.8	436.9	24.2	1.7
524,288	2,621,440	38,380	1,747.6	873.8	25.6	3.5
1,048,576	5,242,880	40,400	3,495.3	1,747.6	26.9	7.0
2,097,152	10,485,760	42,420	6,990.5	3,495.3	28.3	14.0
4,194,304	20,971,520	44,440	13,981.0	6,990.5	29.6	28.0
8,388,608	41,943,040	46,460	27,962.0	13,981.0	31.0	55.9

Figure 2-2. Performance of algorithms X and Y on different computers

While initially X requires fewer operations than Y on problem instances of the same size, once N is 8,192 or larger, Y requires far fewer operations, and it's not even close. The graphs in Figure 2-3 visualize the *crossover* point between 4,096 and 8,192, when Y begins to outperform X in terms of the number of operations required. When you run the exact same implementation of X on two different computers, you can see that X_{fast} (running on C_{fast}) outperforms X_{slow} (running on C_{slow}).

If you found a supercomputer, $C_{fastest}$, that was 500 times faster than C_{slow}, you could eventually find a problem instance size for which the efficient algorithm Y, running on C_{slow}, outperforms the inefficient algorithm X, running on $C_{fastest}$. This is an "apples vs. oranges" comparison in a way, because the programs are running on different computers; nonetheless, in this specific case, the crossover occurs on problem instance sizes between 4,194,304 and 8,388,608. Even with a supercomputer, eventually the more efficient algorithm will outperform it on a slower computer, once the problem instances are large enough.

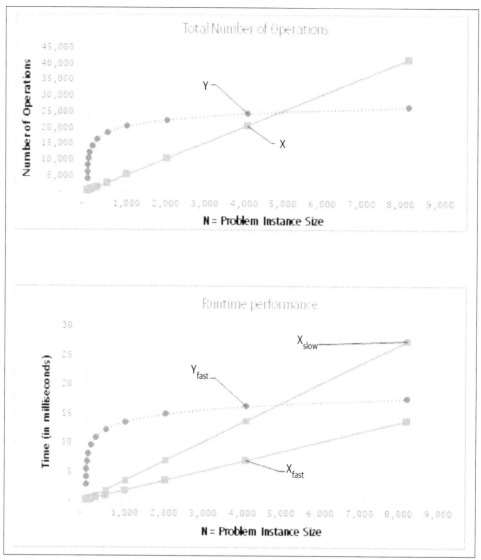

Figure 2-3. Visualizing the numbers from Figure 2-2

You can try to throw advanced computing hardware at a problem, but eventually the more efficient algorithm will be faster for large-enough problem instances.

Computer scientists use a Big O notation to classify algorithms based on the runtime performance on *best case* (or *worst case*) problem instances of size N. The letter O is used because the growth rate of a function is called the "order of a function." For example, the formula $4N^2 + 3N - 5$ is an "order 2" function, also called quadratic, since the largest exponent for N is 2.

To estimate the runtime performance for an algorithm on a problem instance of size N, start by counting the number of operations. It's assumed that each operation takes a fixed amount of time to execute, which turns this count into an estimate for the runtime performance.

T(N) is the time required for an algorithm to process a problem instance of size N. There can be different T(N) defined for *best case* and *worst case* problem instances for the same algorithm. The time unit does not matter (whether milliseconds, or seconds).

S(N) is the storage required for an algorithm to process a problem instance of size N. There can be different S(N) defined for *best case* and *worst case* problem instances for the same algorithm. The space unit does not matter (whether bits or gigabytes).

Counting All Operations

The goal is to estimate the time for an algorithm to process *any problem instance* of size N. Because this estimate must be accurate for all problem instances, try to find a *worst case* problem instance that will force the algorithm to work the most.

First, determine K(N), the count of how many times a *key operation* executes on a *worst case* problem instance of size N. Next, estimate that the number of machine instructions executed in total would be a multiple of this count, that is $c \times K(N)$. This is a safe assumption because modern programming languages can be compiled into tens or hundreds of machine instructions. You don't even have to compute c, but rather you can empirically determine it, based on the individual performance on a computer, as I have done.

The notation clearly classifies the trajectory of the performance (or storage) as a function of N. Each performance class $O(f(N))$ is described by some $f(N)$. The terminology can, at first, be confusing. When classifying the algorithm, you will use a formula represented as a function, f, based on N. We have seen four performance classes:

- $O(N)$ is the *linear complexity class*, where $f(N) = N$.
- $O(N^{1.585})$ is the Karatsuba complexity class, where $f(N) = N^{1.585}$.
- $O(N^2)$ is the *quadratic complexity class*, where $f(N) = N^2$.
- $O(N \log N)$ is the complexity class where $f(N) = N \times \log N$.

To conduct an accurate analysis, you must inspect the source code to see the structure of the algorithm. In the following code example, how many times does the key operation ct = ct + 1 execute?

```
for i in range(100):
    for j in range(N):
        ct = ct + 1
```

The outer i loop executes 100 times, and for each of these loops, the inner j loop executes N times. In total, ct = ct + 1 executes $100 \times N$ times. The total time $T(N)$ to execute the preceding code on a problem instance of size N is smaller than $c \times N$ for some suitably chosen c. If you execute this code on an actual computer, you will be able to determine the exact c. More precisely, using the Big O notation, we can state that the performance of this code is O(N).

Run this code thousands of times on different computing systems, and each time you would be able to compute a different c; this fact remains true and is the reason we can classify the code performance as O(N). There are theoretical details I sidestep in this discussion, but you only need to know that when you have identified a function, $f(N)$, that represents the count of the operations in your algorithm, you have its algorithm classification, $O(f(N))$.

Counting All Bytes

You can perform a similar analysis to determine the space complexity required for an algorithm on a problem instance of size N. When an algorithm dynamically allocates additional storage space, it invariably increases the runtime performance because of the costs associated with dynamic memory management.

The following Python statements require a different amount of space:

- range(N) uses a fixed amount of space because in Python 3, range is a generator that produces the numbers one at a time, without allocating a whole list (as it did in Python 2).
- list(range(N)) constructs a list storing N integers from 0 to N – 1. The size of the required memory grows larger in direct proportion to N.

Quantifying the space for a statement is hard because there is no universally agreed-upon unit for space. Should we count the bytes of memory used? The bits? Does it matter if an integer requires 32 bits of storage or 64 bits of storage? Imagine a future computer that allowed for 128-bit representations of integers. Has the space complexity changed? In Python, sys.getsizeof(...) determines the size in bytes for an object. Python 3 uses generators for range(), which significantly reduces the storage

needs for Python programs. If you type the following statements into a Python interpreter, you will see the corresponding storage requirements:

```
>>> import sys
>>> sys.getsizeof(range(100))
48
>>> sys.getsizeof(range(10000))
48
>>> sys.getsizeof(list(range(100)))
1008
>>> sys.getsizeof(list(range(1000)))
9112
>>> sys.getsizeof(list(range(10000)))
90112
>>> sys.getsizeof(list(range(100000)))
900112
```

These results show that the byte storage for list(range(10000)) is about 100 times
larger than for list(range(100)). And when you review the other numbers, you can
classify this storage requirement as O(N).

In contrast, the number of bytes required for range(100) and range(10000) is identical (48 bytes). Since the storage is constant, we need to introduce another complexity class, known as the *constant complexity class*:

- O(1) is the *constant complexity class*, where f(N) = c for some constant c.

I've covered a lot of theoretical material in this chapter, and it's time to put these concepts to practical use. I now present an optimal searching algorithm in computer science called Binary Array Search. In explaining why it is so efficient, I will introduce a
new complexity class, O(log N).

When One Door Closes, Another One Opens

I have written a sequence of seven different numbers in increasing order from left to
right, and hidden each one behind a door, as shown in Figure 2-4. Try this challenge:
what is the fewest number of doors you need to open—one at a time—to either find a
target value of 643 or prove it is not hidden behind one of these doors? You could
start from the left and open each door—one at a time—until you find 643 or a larger
number (which means it wasn't behind a door in the first place). But with bad luck,
you might have to open all seven doors. This search strategy doesn't take advantage of
the fact that you know the numbers behind the doors are in ascending order. Instead,
you can solve the challenge by opening no more than three doors. Start with door 4
in the middle and open it.

Figure 2-4. Doors of destiny!

The number it was hiding is 173; since you are searching for 643, you can ignore all of the doors to the left of door 4 (since the numbers behind those doors will all be smaller than 173). Now open door 6 to reveal the number 900. OK, so now you know that you can ignore the doors to the right of door 6. Only door 5 can be hiding 643, so open it now to determine whether the original series contained 643. I leave it to your imagination whether the number was behind that door.

If you repeat this process on any ascending list of seven numbers using any target value, you will never need to open more than three doors. Did you notice that $2^3 - 1 = 7$? What if you had 1,000,000 doors covering an ascending list of numbers? Would you accept a $10,000 challenge to determine whether a specific number is hidden behind some door if you are only able to open 20 doors? You should! Since $2^{20} - 1 = 1,048,575$, you can always locate a number in an ascending list of 1,048,575 numbers after opening 20 or fewer doors. Even better, if there were suddenly twice as many doors, 2,097,151 in fact, you would never need to open more than 21 doors to find a number; that's just one additional door to open. That seems astonishingly efficient! You have just discovered Binary Array Search.

Binary Array Search

Binary Array Search is a fundamental algorithm in computer science because of its time complexity. Listing 2-3 contains an implementation that searches for `target` in an ordered list, A.

Listing 2-3. Binary Array Search

```
def binary_array_search(A, target):
  lo = 0
  hi = len(A) - 1          ❶

  while lo <= hi:          ❷
    mid = (lo + hi) // 2   ❸

    if target < A[mid]:    ❹
      hi = mid-1
    elif target > A[mid]:  ❺
      lo = mid+1
    else:
      return True          ❻

  return False             ❼
```

❶   Set lo and hi to be inclusive within list index positions of 0 and len(A)-1.

❷   Continue as long as there is at least one value to explore.

❸   Find midpoint value, A[mid], of remaining range A[lo .. hi].

❹   If target is smaller than A[mid], continue looking to the *left* of mid.

❺   If target is larger than A[mid], continue looking to the *right* of mid.

❻   If target is found, return True.

❼   Once lo is greater than hi, there are no values remaining to search. Report that target is not in A.

Initially lo and hi are set to the lowest and highest indices of A. While there is a sublist to explore, find the midpoint, mid, using integer division. If A[mid] is target, your search is over; otherwise you have learned whether to repeat the search in the sublist to the left, A[lo .. mid-1], or to the right, A[mid+1 .. hi].

The notation A[lo .. mid] means the sublist from lo up to and including mid. If lo > mid, then the sublist is empty.

This algorithm determines whether a value exists in a sorted list of N values. As the loop iterates, eventually either the `target` will be found or `hi` crosses over to become smaller than `lo`, which ends the loop.

Almost as Easy as π

Consider using Binary Array Search to find a target value of 53 in the list shown in Figure 2-5. First, set `lo` and `hi` to the boundary index positions of A. In the `while` loop, `mid` is computed. Since `A[mid]` is 19—which is smaller than the target, 53—the code takes the `elif` case, setting `lo` to `mid + 1` to refine the search on the sublist `A[mid+1 .. hi]`. The grayed-out values are no longer in consideration. The size of the sublist being explored after this iteration is reduced by half (from 7 values to 3).

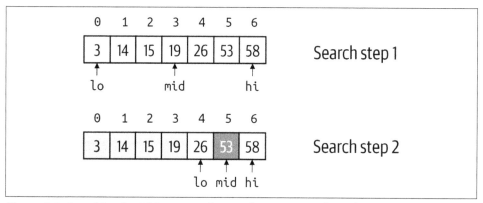

Figure 2-5. Searching for 53 in a sorted array that contains the value

In the second pass through the `while` loop, `mid` is recomputed, and it turns out that `A[mid]` is 53, which is the target value, so the function returns `True`.

 Before conducting Binary Array Search, is it worth checking that `target ≥ A[0]` and `target ≤ A[-1]`? Doing so would prevent a fruitless search for a target that couldn't possibly be present in an ordered list. The short answer is no. This adds up to *two* comparisons to every search, which are unnecessary if the searched-for values are always within the range of the extreme values in A.

Now let's search for a value that is not in the list. To search for the target value of 17 in Figure 2-6, initialize lo and hi as before. A[mid] is 19, which is larger than the target, 17, so take the if case and focus the search on A[lo .. mid-1]. The grayed-out values are no longer in consideration. The target, 17, is greater than A[mid] = 14, so take the elif case and try searching A[mid+1 .. hi].

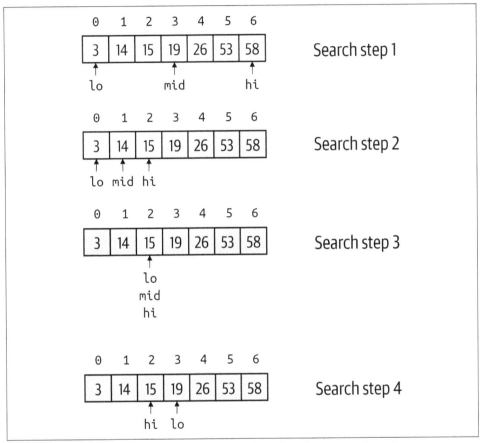

Figure 2-6. Searching for 17 in a sorted array that doesn't contain the value

In the third time through the while loop, A[mid] is 15, which is smaller than the target value of 17. Once again, take the elif case, which sets lo to be larger than hi; this is the "crossover," shown at the bottom of Figure 2-6. The condition of the while loop is false, and so False is returned, declaring that A does not contain the target value.

Two Birds with One Stone

What if you want to know the exact location of `target` in `A`, instead of just confirming that `target` is contained in `A`? Binary Array Search currently returns `True` or `False`. Modify the code, as shown in Listing 2-4, to return the index position, `mid`, where `target` is found.

Listing 2-4. Return location of `target` *in* `A`

```
def binary_array_search(A, target):
  lo = 0
  hi = len(A) - 1

  while lo <= hi:
    mid = (lo + hi) // 2

    if target < A[mid]:
      hi = mid-1
    elif target > A[mid]:
      lo = mid+1
    else:
      return mid            ❶

  return -(lo+1)            ❷
```

❶ Return the value of `mid` since that is the location of `target`.

❷ Alert caller that `target` doesn't exist by returning the negative of `lo + 1`.

What should be returned when `target` is not in `A`? You could just return –1 (which is an invalid index location), but there is an opportunity to return more information. What if we could tell the caller "`target` is not in `A`, but if you wanted to insert `target` into `A`, it would go in *this* location"?

Look at Figure 2-6 again. When searching for a target value of 17 (which doesn't exist in `A`), the final value of `lo` is actually where 17 *would be inserted*. You could return –`lo` as the result, and this would work for all index locations *except for the first one*, which is zero. Instead return the negation of (`lo` + 1). The calling function that receives a negative value, `x`, has learned that `target` would be placed at location `-(x + 1)`. When a nonnegative value is returned, that is the location of `target` in `A`.

One final optimization remains. In Listing 2-4, there are two comparisons between `target` and `A[mid]`. When both values are numeric, Listing 2-5 shows how to compute their difference *just once* instead of invoking this key operation twice; this also ensures you only access `A[mid]` once.

Listing 2-5. Optimization that requires just a single value comparison

```
diff = target - A[mid]
if diff < 0:
  hi = mid-1
elif diff > 0:
  lo = mid+1
else:
  return mid
```

If `target` is smaller than `A[mid]`, then `diff` < 0, which is equivalent logically to checking whether `target` < `A[mid]`. If `diff` is positive, then you know `target` was greater than `A[mid]`. Even when the values are not numeric, some programming languages offer a `compareTo()` function that returns a negative number, zero, or a positive number based on the relative ordering of two values. Using this operation leads to more efficient code when comparison operations are costly.

 If the values in a list appear in descending order, you can still use Binary Array Search—just switch the way `lo` and `hi` are updated in the `while` loop.

How efficient is Binary Array Search on a problem instance of size N? To answer this question, I have to compute, in the *worst case*, how many times the `while` loop is forced to execute. The mathematical concept of logarithms will tell us the answer.[5]

To see how logarithms work, consider this question: how many times do you need to double the number 1 until the value equals 33,554,432? Well, you could start computing this manually: 1, 2, 4, 8, 16, and so on, but this is truly tedious. Mathematically, you are looking for a value, x, such that 2^x = 33,554,432.

Note that 2^x involves *exponentiation* of a base (the value 2) and an exponent (the value x). A logarithm is the opposite of exponentiation, in the same way that division is the opposite of multiplication. To find an x such that 2^x = 33,554,432, compute $\log_2$ (33,554,432) using a base of 2, resulting in the value 25.0. If you type the equation 2^{25} into a calculator, the result is 33,554,432.

This computation also answers the question of how many times you can divide 2 into 33,554,432. You get to 1 after 25 divisions. `log()` computes a floating point result; for example, $\log_2(137)$ is about 7.098032. This makes sense since 2^7 = 128, and 137 would require a slightly higher exponent for base 2.

---

5 It is purely a coincidence that the word *logarithm* is an anagram of *algorithm*.

The Binary Array Search algorithm will repeat the `while` loop as long as `lo ≤ hi`, or in other words, while there are values to search. The first iteration starts with N values to search, and in the second iteration, this is reduced to no more than N/2—if N is odd it is (N – 1)/2. To determine the maximum number of successive iterations, you need to know how many times you can divide N by 2 until you reach 1. This quantity is exactly $k = \log_2(N)$, so the total number of times through the `while` loop is $1 + k$, counting 1 for the first time for N values plus k for the successive iterations. Because `log()` can return a floating point value, and we need an integer number for the number of iterations, use the `floor(x)` mathematical operation, which computes the largest integer that is smaller than or equal to x.

No handheld calculator has a button to compute $\log_2(X)$—most calculator apps don't either. Don't panic! You can always compute $\log_2$ easily. For example, $\log_2(16) = 4$. On your calculator, enter 16 and then press the `log` button (which is either in base 10 or the natural base e). Your display should read something awful like 1.20411998. Now press the / (divide) button, press the 2 button, and finally press the `log` button again. Your display should read 0.301029995. Now that all hope seems lost, press the equals button. Magically the value 4 appears. This sequence of operations demonstrates that $\log_2(X) = \log_{10}(X)/\log_{10}(2)$.

For Binary Array Search, the `while` loop iterates no more than $\text{floor}(\log_2(N)) + 1$ times. This behavior is truly extraordinary! With one million values in sorted order, you can locate any value in just 20 passes through the `while` loop.

To provide some quick evidence for this formula, count the number of iterations through the `while` loop for problem instances whose size, N, ranges from 8 to 15: you only need 4 in all cases. For example, starting with 15 values in the first iteration, the second iteration explores a sublist with 7 values, the third iteration explores a sublist with 3 values, and the fourth and final iteration explores a sublist with just 1 value. If you started with 10 values, the number of explored values in each iteration would be 10 → 5 → 2 → 1, which also means four iterations in total.

The experience with Binary Array Search leads to a new complexity class, O(log N), called the *logarithmic complexity class*, where $f(N) = \log(N)$.

To summarize, when you analyze an algorithm and state that its time complexity is O(log N), you are claiming that once the problem instance size is larger than some threshold size, the runtime performance, T(N), of the algorithm is always smaller than $c \times \log(N)$ for some constant, c. Your claim is correct *if you can't make this claim with another complexity class* of lower complexity.

All complexity classes are arranged in order of dominance, as shown in Figure 2-7.

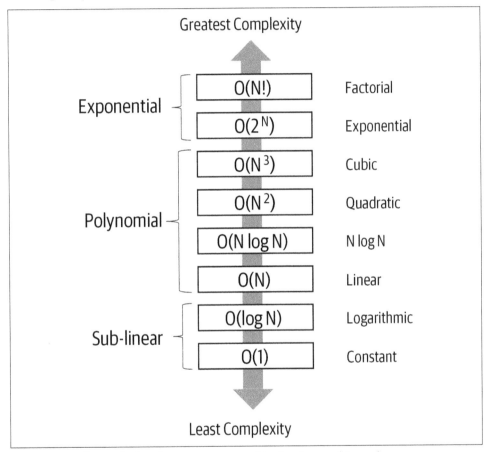

Figure 2-7. *All complexity classes are arranged in dominance hierarchy*

While there are an infinite number of complexity classes, the eight classes in this figure are the most commonly used ones. The constant time, O(1), has the least amount of complexity and reflects constant work that is independent of the size of the problem instance. The next-higher complexity class, O(log N), is logarithmic, and you have seen how Binary Array Search falls into this category. Both of these classes are *sub-linear* and result in extremely efficient algorithms.

The linear, O(N), complexity class means complexity is directly proportional to the size of the problem instance. A series of polynomial classes are all of increasing complexity—$O(N^2)$, $O(N^3)$, and so on—up to any fixed constant, $O(N^c)$. Sandwiched between O(N) and $O(N^2)$ is O(N log N), which is often identified as the ideal complexity class for algorithm designers.

This dominance hierarchy can also help identify how to classify an algorithm when there is a mixed combination of time complexities. For example, if an algorithm contains two substeps—the first with time complexity of O(N log N) and the second with $O(N^2)$—what is its overall complexity? The overall classification of the algorithm is $O(N^2)$ because the complexity of substep 2 is the dominant impact on the overall complexity. In practical terms, if you have modeled T(N) for an algorithm to be $5N^2 + 10,000,000 \times N \times \log(N)$, then T(N) has complexity of $O(N^2)$.

The final two complexity classes—exponential and factorial—are inefficient, and algorithms with these time complexities can only solve very small problem instances. Check out the challenge exercises at the end of the chapter that address these complexity classes.

Pulling It All Together

Table 2-4 presents the computation of f(N) for each of the designated complexity classes. Imagine that each of these numbers represents the estimated time in seconds for an algorithm with the rated time complexity (in a column) to process a problem instance of size N (as identified by the row). 4,096 is about one hour and eight minutes, so in just over an hour of computation time, you could likely solve:

- An O(1) algorithm, since its performance is independent of the problem instance size
- An O(log N) algorithm for problem instances of size 2^{4096} or smaller
- An O(N) algorithm for problem instances of 4,096 or smaller
- An O(N log N) algorithm for problem instances of size 462 or smaller
- An $O(N^2)$ algorithm for problem instances of size 64 or smaller
- An $O(N^3)$ algorithm for problem instances of size 16 or smaller
- An $O(2^N)$ algorithm for problem instances of size 12 or smaller
- An O(N!) algorithm for problem instances of size 7 or smaller

Table 2-4. Growth of different computations

N	log(N)	N	N log N	N²	N³	2ᴺ	N!
2	1	2	2	4	8	4	2
4	2	4	8	16	64	16	24
8	3	8	24	64	512	256	40,320
16	4	16	64	256	4,096	65,536	2.1×10^{13}
32	5	32	160	1,024	32,768	4.3×10^{9}	2.6×10^{35}
64	6	64	384	4,096	262,114	1.8×10^{19}	1.3×10^{89}
128	7	128	896	16,384	2,097,152	3.4×10^{38}	∞
256	8	256	2,048	65,536	16,777,216	1.2×10^{77}	∞
512	9	512	4,608	262,144	1.3×10^{8}	∞	∞
1,024	10	1,024	10,240	1,048,576	1.1×10^{9}	∞	∞
2,048	11	2,048	22,528	4,194,304	8.6×10^{9}	∞	∞

The reason to investigate algorithms with the lowest complexity rating is because the problems you want to solve are simply too large on even the fastest computer. With higher complexity classes, the time to solve even small problems essentially is infinite (as Figure 2-8 shows).

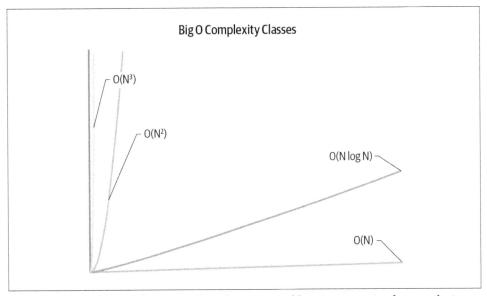

Figure 2-8. Runtime performance plotted against problem instance size for complexity classes

A common way to visualize these extremely large numbers is shown in Figure 2-8. The x-axis represents the size of the problem instance being solved. The y-axis represents the total estimated runtime performance for one of the algorithms labeled in the chart. As the complexity of the algorithm increases, the size of the problem instance that can be solved "in reasonable time" decreases.

Consider the following, more complex scenarios:

- If someone classifies an algorithm as $O(N^2 + N)$, how should you respond? The dominance hierarchy in Figure 2-7 shows that N^2 has a greater complexity than N, and so this can be simplified to be $O(N^2)$. Similarly, $O(2^N + N^8)$ would be simplified to become $O(2^N)$.

- If an algorithm is classified as $O(50 \times N^3)$, you can simplify this to become $O(N^3)$ because multiplicative constants can be eliminated.

- Sometimes the behavior of an algorithm can depend on properties other than just the size of the problem instance, N. For example, consider an algorithm that processes N numeric values, with a primary task whose runtime performance is directly proportional to N. Now assume this algorithm has a subtask that processes all *even values* in the problem instance. The runtime performance of this subtask is directly proportional to E^2, where E represents the number of even values. You might want to carefully specify that the runtime performance of the algorithm is $O(N + E^2)$. If, for example, you could eliminate all even values from the input set, this performance would be rated as $O(N)$, and that might be noteworthy. Naturally, in the *worst case* where all numbers are even, then the overall classification of the algorithm would become $O(N^2)$ since $E \leq N$.

Curve Fitting Versus Lower and Upper Bounds

The curve_fit() function provided by SciPy uses a nonlinear, least squares method to fit a model function, f, to existing data. I use data based on different problem instances of size N and the respective runtime performance of an algorithm in solving those instances. The result of curve_fit()—as you have seen in this chapter—are coefficients to use with a model function to predict future runtime performance. With these coefficients applied to f, the resulting model *minimizes the sum of the square of the errors* between the actual data and the predicted value from the model.

These are useful to get a ballpark estimate of the inner behavior of an algorithm's implementation on specific problem instances. By itself, this model is neither a proven upper bound or lower bound regarding the complexity of an algorithm. You need to review the algorithm's implementation to develop a model that counts the number of key operations, which directly affects the runtime performance of the algorithm.

When you have an accurate model, f(N), that reflects the count of the key operations in the *worst case* for an algorithm on a problem instance of size N, then you have classified O(f(N)) in the *worst case*. This is the upper bound. A corresponding lower bound can be similarly derived to model the effort that the algorithm must at least expend in the *worst case*. The notation Ω(f(N)) is used to describe the classification of the lower bound of an algorithm.[6]

In our earlier discussion of Binary Array Search, I demonstrated that the while loop iterates no more than floor(log₂(N)) + 1 times. This means, in the *worst case*, use f(N) = log(N) to formally classify Binary Array Search as O(log N). What is the *best case* for Binary Array Search? If the target value is found in A[mid], the function returns after just one pass through the while loop. Since this is a constant *independent of the size of the problem instance*, this means that Binary Array Search is classified as O(1) in the *best case*. The big O notation can be used for both *best case* and *worst case* analysis, although many programmers assume it is meant only for *worst case*.

 You may occasionally see the time complexity for an algorithm rated as Θ(N log N) using the capitalized Greek symbol theta. This notation is typically used to analyze the *average case* for an algorithm. This means that the upper bound is O(N log N), and the lower bound is Ω(N log N). This is known, mathematically, as a *tight bound* and provides the best evidence that the runtime performance of the algorithm is quite predictable.

Summary

We have covered a lot of ground in these first two chapters, but there is so much more you can learn about analyzing algorithms. I presented a number of examples that describe the way algorithms behave *independent of how they are implemented*. In the mid-20th century, while researchers were discovering new algorithms, advances in computing technology dramatically improved the very performance of the computers executing these algorithms. Asymptotic analysis provides the foundation for independently assessing the performance of algorithms in a way that eliminates any dependence on the computing platform. I defined several time (or storage) complexity classes, visualized in Figure 2-8, to explain the behavior of an algorithm in terms of the size of the problem instances. These complexity classes appear throughout the book as a notation to quickly summarize an algorithm's behavior.

---

6 Ω is the capitalized Greek character omega.

Challenge Exercises

1. Rate the time complexity of each code fragment in Table 2-5.

   Table 2-5. Code fragments to analyze

   ```
   Fragment-1  for i in range(100):
                   for j in range(N):
                     for k in range(10000):
                       ...

   Fragment-2  for i in range(N):
                   for j in range(N):
                     for k in range(100):
                       ...

   Fragment-3  for i in range(0,N,2):
                   for j in range(0,N,2):
                     ...

   Fragment-4  while N > 1:
                   ...
                   N = N // 2

   Fragment-5  for i in range(2,N,3):
                   for j in range(3,N,2):
                     ...
   ```

2. Use the techniques described in this chapter to model the value of `ct` returned by the `f4` function in Listing 2-6.

   Listing 2-6. Sample function to analyze

   ```
   def f4(N):
     ct = 1
     while N >= 2:
       ct = ct + 1
       N = N ** 0.5
     return ct
   ```

   You will find that none of the models used in this chapter is accurate. Instead, develop one based on $a \times \log(\log(N))$, in base 2. Generate a table up to $N = 2^{50}$ containing actual results as compared to the model. An algorithm with this behavior would be classified as $O(\log(\log(N)))$.

3. One way to sort a list of values is to generate each permutation until you find one that is sorted, as shown in Listing 2-7.

Listing 2-7. Code to generate permutations from a list

```
from itertools import permutations
from scipy.special import factorial

def factorial_model(n, a):
  return a*factorial(n)

def check_sorted(a):
  for i, val in enumerate(a):
    if i > 0 and val < a[i-1]:
      return False
  return True

def permutation_sort(A):
  for attempt in permutations(A):
    if check_sorted(attempt):
      A[:] = attempt[:]       # copy back into A
      return
```

Generate a table of results for sorting a *worst case* problem instance (i.e., the values are in descending order) of up to 12 elements using permutation_sort(). Use the factorial_model() to curve fit the preliminary results and see how accurate the model is in predicting runtime performance. Based on these results, what is your estimate (in years) for the runtime performance on a *worst case* problem instance of size 20?

4. Generate empirical evidence on 50,000 random trials of Binary Array Search for N in the range of 2^5 through 2^{21}. Each trial should use random.sample() to randomly select N values from the range 0 .. 4N and place these values in sorted order. Then each trial should search for a random target value in the same range.

   Using the results I have outlined in this chapter, use curve_fit() to develop a log N model that models the results of runtime performance for N in the range 2^5 through 2^{12}. Determine the *threshold problem instance size* above which the behavior stabilizes. Create a visual plot of the data to see whether the computed model accurately models the empirical data.

5. We are normally concerned with time complexity, but consider the sorting algorithm in Listing 2-8:

Listing 2-8. Code to sort by repeatedly removing maximum value from list

```
def max_sort(A):
  result = []
  while len(A) > 1:
    index_max = max(range(len(A)), key=A.__getitem__)
    result.insert(0, A[index_max])
    A = list(A[:index_max]) + list(A[index_max+1:])
  return A + result
```

Using the results I have outlined in this chapter, assess the storage complexity of `max_sort`.

6. A *galactic algorithm* is an algorithm whose time complexity is better than any known algorithm when the problem instance size is "sufficiently large." For example, the N-digit multiplication algorithm (published November 2020) by David Harvey and Joris Van Der Hoeven has $O(N \log N)$ runtime performance, once N is larger than 2^Z, where Z is 1729^{12}; this exponent, Z, is already an astronomically large number, about 7×10^{38}. Consider now raising 2 to this incredibly large number! Do some research on other galactic algorithms. While these algorithms are not practical, they do offer hope that breakthroughs are possible on some really challenging problems.

7. Table 2-1 contains three rows of performance measurements on three data sets of different sizes. If you only had two rows of performance measurements, would it be possible to predict the performance of a quadratic time algorithm? In general, if you have K rows of performance measurements, what is the highest polynomial you can effectively use in a model?

## CHAPTER 3

Better Living Through Better Hashing

In this chapter, you will learn:

- How to store (key, value) pairs in a *symbol table* and retrieve values associated with a key.[1]
- How to use an array to store (key, value) pairs for efficient search, if the size of the array is sufficiently large, compared to the number of pairs stored.
- How to use an array of linked lists to store (key, value) pairs to support the extra ability to remove a key.
- How to resize a symbol table to remain efficient.
- *Amortized analysis* to determine the *average* runtime performance when the behavior of an operation can change with successive invocations.
- How *geometric resizing* reduces the frequency of the costly resize operation, meaning that put() has amortized $O(1)$ performance on average.
- How a computational *hash function* can uniformly distribute key values, which ensures the efficiency of the symbol table implementation.

Associating Values with Keys

Instead of just storing values, you might need to store a collection of (key, value) pairs to associate values with specific keys. This is known as a *symbol table* data type, which lets you find the associated value given just its key. Hashing offers an efficient

---

1 Throughout this chapter, the notation (key, value) represents a pair of information considered as a single unit.

alternative to manually searching through a collection from start to finish just to find a (key, value) pair. It outperforms the search algorithms I covered earlier. A symbol table can be efficient even while allowing for keys (and their values) to be removed. You give up the ability to retrieve all keys in a specific order, for example, in ascending order, but the resulting symbol table provides optimal performance for retrieving or storing the value associated with any individual key.

Let's say you want to write a function `print_month(month, year)` to print a calendar for any month and year. For example, `print_month('February', 2024)` would output the following:

```
    February 2024
Su Mo Tu We Th Fr Sa
                1  2  3
 4  5  6  7  8  9 10
11 12 13 14 15 16 17
18 19 20 21 22 23 24
25 26 27 28 29
```

What information would you need? You need the weekday for the first day of that month in that year (above it is Thursday), and you need to know that February has 28 days (or 29 in a leap year, such as 2024). You might use a fixed array, `month_length`, with values that record the length of each month in days for the year:

```
month_length = [ 31, 28, 31, 30, 31, 30, 31, 31, 30, 31, 30, 31]
```

January, the first month, has 31 days, so `month_length[0]` = 31. February is the next month, with 28 days, so the next value in the list is 28. The final value in `month_length` is 31, since December, the last month of the year, has 31 days.

Given what I have presented in this book so far, you might choose to store a `key_array` of the same length and search it for the month to determine the corresponding value in `month_length`; the following code prints `February has 28 days`:

```
key_array    = [ 'January', 'February', 'March', 'April', 'May', 'June', 'July',
                 'August', 'September', 'October', 'November', 'December' ]

idx = key_array.index('February')
print('February has', month_length[idx], 'days')
```

While the preceding code snippet will work, in the *worst case*, the key you are looking for is either the last one in the list (December) or an invalid month name, which means you have to inspect all values in the array. This means the time to locate the associated value for a key is *directly proportional to the number of keys stored*. If there were hundreds of thousands of (key, value) pairs, this approach would rapidly become so inefficient as to be unusable. Given this starting point, you should try to implement `print_month()`; for the record, Listing 3-1 contains my implementation, which uses the `datetime` and `calendar` Python modules.

Listing 3-1. Code to print readable calendar for any month and year

```
from datetime import date
import calendar

def print_month(month, year):
  idx = key_array.index(month)                ❶
  day = 1

  wd = date(year,idx + 1,day).weekday()       ❷
  wd = (wd + 1) % 7                            ❸
  end = month_length[idx]                      ❹
  if calendar.isleap(year) and idx == 1:       ❺
    end += 1

  print('{} {}'.format(month,year).center(20))
  print('Su Mo Tu We Th Fr Sa')
  print('   ' * wd, end='')                    ❻
  while day <= end:
    print('{:2d} '.format(day), end='')
    wd = (wd + 1) % 7                           ❼
    day += 1
    if wd == 0: print()                         ❽
  print()
```

❶ Find index to use for `month_length`, an integer from 0 to 11.

❷ Returns weekday for first day of given month, using 0 for Monday. Note `date()` uses `idx + 1` since its month argument must be an integer from 1 to 12.

❸ Adjust to have Sunday be the weekday with a value of 0, instead of Monday.

❹ Determine length of the month corresponding to input parameter.

❺ In a leap year, February (index 1 when counting from 0) has 29 days.

❻ Add spaces in first week to start day 1 at right indentation.

❼ Advance day and weekday for next day.

❽ Insert line break before Sunday to start a new line.

For the task of finding the associated value for a given key in a collection of N (key, value) pairs, I judge efficiency by counting the number of times any array index is accessed. A function searching for a string in `key_array` could inspect up to N array index positions, so that function's performance is O(N).

 Python provides a `list` data structure instead of an array. While lists can dynamically grow in size, in this chapter, I continue to use the term *array* since I do not take advantage of this capability.

Python has a built-in `dict` type (an abbreviation of *dictionary*) that associates values with keys. In the following, `days_in_month` is a `dict` that associates an integer *value* (representing the length of that month) with a string *key* containing the capitalized English name:

```
days_in_month = { 'January'   : 31,  'February'  : 28,  'March'     : 31,
                  'April'     : 30,  'May'       : 31,  'June'      : 30,
                  'July'      : 31,  'August'    : 31,  'September' : 30,
                  'October'   : 31,  'November'  : 30,  'December'  : 31 }
```

The following code prints `April has 30 days`:

```
print('April has', days_in_month['April'], 'days')
```

The `dict` type can locate a key with an average performance of O(1), independent of the number of (key, value) pairs it stores. This is an amazing accomplishment, like a magician pulling a rabbit from a hat! In Chapter 8, I provide more details about `dict`. For now, let's see how it works. The following code provides some mathematical intuition as to how this happens. The key idea is to turn a string into a number.

Try this: consider the letter `'a'` to be the value 0, `'b'` to be 1, and so on through `'z'` = 25. You can then consider `'June'` to be a number in base 26, which represents the base 10 value j × 17,576 + u × 676 + n × 26 + e = 172,046 in base 10.[2] This computation can also be written as 26 × (26 × (26 × j + u) + n) + e, which reflects the structure of the `base26()` method shown in Listing 3-2.

Listing 3-2. Convert word to an integer assuming base 26

```
def base26(w):
  val = 0
  for ch in w.lower():          ❶
    next_digit = ord(ch) - ord('a')   ❷
    val = 26*val + next_digit         ❸
  return val
```

---

2 Note that $26^3 = 17,576$.

❶ Convert all characters to lowercase.

❷ Compute digit in next position.

❸ Accumulate total and return.

base26() uses the ord() function to convert a single character (such as 'a') into its integer ASCII representation.[3] ASCII codes are ordered alphabetically, so ord('a') = 97, ord('e') = 101 and ord('z') = 122. To find the value associated with 'e', simply compute ord('e') - ord('a') to determine that 'e' represents the value 4.

When computing base26() values for a string, the resulting numbers quickly grow large: 'June' computes to the value 172,046, while 'January' computes to 2,786,534,658.

How can these numbers be cut down to a more manageable size? You might know about the *modulo* operator % provided by most programming languages. This operator returns the integer remainder when dividing a large integer (i.e., the base26(month) computation) by a much smaller integer.

You have likely used modulo in real life without knowing it. For example, if it is currently 11:00 a.m., what time of day will it be in 50 hours? You know in 24 hours it will once again be 11:00 a.m. (on the next day), and after 48 hours it will again be 11:00 a.m. (on the second day). This leaves only two hours to account for, which tells you that in 50 hours it will be 1:00 p.m. Mathematically speaking, 50 % 24 = 2. In other words, you cannot evenly divide 50 by 24 since you will have 2 left over. When N and M are positive integers, N % M is guaranteed to be an integer from 0 to M – 1.

With some experimentation, I discovered that base26(m) % 34 computes a different integer for each of the twelve English month names; for example, base26('August') computes to 9,258,983, and you can confirm that 9,258,983 % 34 = 1. If you create a single array containing 34 values, as shown in Figure 3-1, then *independently of the number of (key, value) pairs*, you can determine the number of days in a month by computing its associated index into day_array. This means that August has 31 days; a similar computation for February shows it has 28 days.

---

3 Originally based on the English alphabet, ASCII encodes 128 characters (typically those found on a typewriter keyboard) into seven-bit integers. Capitalization matters! ord('A') = 65, for example, but ord('a') = 97.

Take a moment to reflect on what I have accomplished. Instead of iteratively searching for a key in an array, I can now perform a simple computation *on the key itself* to compute an index position that contains its associated value. The time it takes to perform this computation is *independent of the number of keys* stored. This is a major step forward!

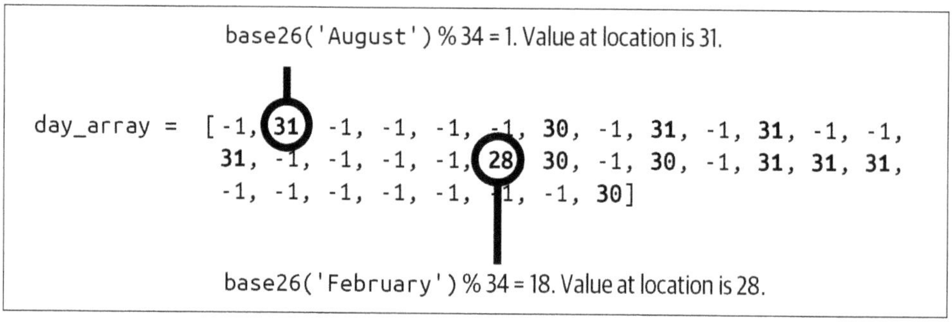

Figure 3-1. Array containing month lengths interspersed with unneeded −1 values

But this was a lot of work. I had to (a) craft a special formula to compute unique index positions, and (b) create an array containing 34 integers, only 12 of which are relevant (meaning that more than half of the values are wasted). In this example, N equals 12, and the amount of dedicated storage, M, equals 34.

Given any string s, if the value in `day_array` at index position `base26(s) % 34` is −1, you know s is an invalid month. This is a good feature. You might be tempted to confirm that a given string, s, is a valid month whenever `day_array[base26(s)%34] > 0`, but this would be a mistake. For example, the string `'abbreviated'` computes to the same index position as `'March'`, which might mistakenly tell you that `'abbreviated'` is a valid month! This is a bad feature, but I now show how to overcome this issue.

Hash Functions and Hash Codes

`base26()` is an example of a *hash function* that maps keys of arbitrary size to a fixed-size *hash code* value, such as a 32-bit or 64-bit integer. A 32-bit integer is a value from −2,147,483,648 to 2,147,483,647, while a 64-bit integer is a value from −9,223,372,036,854,775,808 to 9,223,372,036,854,775,807. As you can see, a hash code can be negative.

Mathematicians have been studying hashing for decades, developing computations to convert structured data into fixed-size integers. Programmers can take advantage of their contributions since most programming languages have built-in support for computing a hash code for arbitrary data. Python provides such a `hash()` method for immutable objects.

The only necessary property of a hash function is that two objects that are equal to each other *must compute to the same hash code*; it cannot simply be a random number. When hashing an immutable object (such as a string), the computed hash code is usually stored to reduce overall computation.

Hash functions do not have to compute a *unique hash* for each key; this is too great a computational challenge (but see the section on perfect hashing at the end of this chapter). Instead, the expression hash(key) % M uses the modulo operation to compute a value guaranteed to be an integer from 0 to M − 1.

Table 3-1 lists the 64-bit hash() value for a few string keys and the corresponding hash code modulo expressions. Mathematically, the probability that two keys have the exact same hash() value is vanishingly small.[4] There are two potential hash code collisions; both *smell* and *rose* have a hash code of 6, while both *name* and *would* have a hash code of 10. You should expect that two different strings could have the exact same hash code value.

Table 3-1. Example hash() and hash code expressions for a table of size 15

key	hash(key)	hash(key) % 15
a	−7,995,026,502,214,370,901	9
rose	−3,472,549,068,324,789,234	6
by	−6,858,448,964,350,309,867	8
any	2,052,802,038,296,058,508	13
other	4,741,009,700,354,549,189	14
name	−7,640,325,309,337,162,460	10
would	274,614,957,872,931,580	10
smell	7,616,223,937,239,278,946	6
as	−7,478,160,235,253,182,488	12
sweet	8,704,203,633,020,415,510	0

If I use hash(key) % M to compute the hash code for key, then M must be at least as large as the number of expected keys to make room to store all associated values.[5]

---

4 Java computes 32-bit hash function values, and sometimes two string keys have the exact same hash() value; for example, both *misused* and *horsemints* hash to 1,069,518,484.

5 In Java, if hash(key) is negative, then the % operator will return a negative number, so the formula must be (key.hashCode() & 0x7fffffff) % M to first convert the negative hash computation into a positive integer before computing modulo M.

Currently, Java and Python 2 compute a predictable hash code for strings. In Python 3, the default hash() code values for strings are "salted" with an unpredictable random value. Although they remain constant within an individual Python process, they are not predictable between repeated invocations of Python as a cyberse-curity measure. Specifically, if a hacker can generate keys that produce specific hash code values (thus violating the uniform distribution of keys), the performance of the hashtable defined in this chapter degrades to O(N), leading to denial-of-service attacks.[6]

A Hashtable Structure for (Key, Value) Pairs

The following Entry structure stores a (key, value) pair:

```
class Entry:
    def __init__(self, k, v):
        self.key = k
        self.value = v
```

Listing 3-3 defines a Hashtable class that constructs a table array capable of storing up to M Entry objects. Each of these M index positions in the array is called a *bucket*. For our first attempt, either a bucket is empty or it contains a single Entry object.

Listing 3-3. Ineffective hashtable implementation

```
class Hashtable:
  def __init__(self, M=10):
    self.table = [None] * M        ❶
    self.M = M

  def get(self, k):                ❷
    hc = hash(k) % self.M
    return self.table[hc].value if self.table[hc] else None

  def put(self, k, v):             ❸
    hc = hash(k) % self.M
    entry = self.table[hc]
    if entry:
      if entry.key == k:
        entry.value = v
      else:                        ❹
        raise RuntimeError('Key Collision: {} and {}'.format(k, entry.key))
    else:
      self.table[hc] = Entry(k, v)
```

---

6 See *https://oreil.ly/C4V0W* to learn more and try the challenge exercise at the end of this chapter.

❶ Allocate a `table` to hold M `Entry` objects.

❷ The `get()` function locates the `entry` associated with the hash code for k and returns its value, if present.

❸ The `put()` function locates the `entry` associated with the hash code for k, if present, and overwrites its value; otherwise it stores a new entry.

❹ A collision occurs when two different keys map to the same bucket identified by its hash code value.

You can use this hashtable as follows:

```
table = Hashtable(1000)
table.put('April', 30)
table.put('May', 31)
table.put('September', 30)

print(table.get('August'))      # Miss: should print None since not present
print(table.get('September'))   # Hit: should print 30
```

If everything works properly, three `Entry` objects will be created for these three (key, value) pairs in an array that has room for 1,000 objects. The performance of `put()` and `get()` will be independent of the number of `Entry` objects in the array, so each action can be considered to have constant time performance, or O(1).

A *miss* occurs when `get(key)` fails to find an `Entry` in the bucket identified by the hash code for `key`. A *hit* occurs when `get(key)` finds an `Entry` whose key matches key. These are both normal behaviors. However, *there is still no strategy to resolve collisions* when the hash codes for two (or more) keys compute to the same bucket. If you don't resolve collisions, then a `put()` could overwrite an existing `Entry` for a different key, causing the `Hashtable` to lose keys—this must be prevented.

Detecting and Resolving Collisions with Linear Probing

It may happen that two entries, e1 = (key1, val1) and e2 = (key2, val2), share the same hash code for their respective keys *even though key1 is different from key2*. In Table 3-1, both 'name' and 'would' have the same hash code of 10. Assume that e1 is put in the `Hashtable` first. You will encounter a *hash collision* later when you try to put e2 in the same `Hashtable`: this is because the identified bucket is nonempty with an `Entry` whose key, key1, is different from key2, the key for e2. If you can't resolve these collisions, you won't be able to store both e1 and e2 in the same `Hashtable`.

Open addressing resolves hash collisions by *probing* (or searching) through alternative locations in `table` when `put()` encounters a collision. One common approach, called *linear probing*, has `put()` incrementally check higher index positions in `table` *for the next available empty bucket* if the one designated by the hash code contains a different entry; if the end of the array is reached without finding an empty bucket, `put()` continues to search through `table` starting from index position 0. This search is guaranteed to succeed *because I make sure to always keep one bucket empty*; an attempt to put an entry into the last remaining empty bucket will fail with a runtime error signifying the hashtable is full.

You might wonder why this approach will even work. Since entries can be inserted into an index position different from their `hash(key)`, how will you find them later? First, you can see that entries are never removed from `Hashtable`, only inserted. Second, as more `Entry` objects are inserted into buckets in the `table` structure, long runs of nonempty buckets appear in `table`. An `Entry`, `e1`, could now be in a different location, anywhere from `hash(e1.key)` or, searching to the right, *until the next empty bucket is found* (wrapping around as necessary).

To demonstrate collision resolution, let's now add five entries into a `Hashtable`, with M = 7 buckets in Figure 3-2 (which only shows their keys). Buckets shaded in gray are empty. Key 20 is inserted into `table[6]`, since 20 % 7 = 6; similarly, 15 is inserted into `table[1]`, and 5 is inserted into `table[5]`.

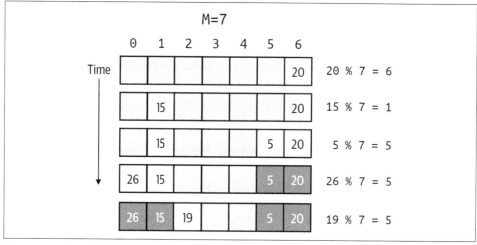

Figure 3-2. Structure of Hashtable storage after adding five (key, value) entries

Adding an entry with key 26 causes a collision since `table[5]` is already occupied (by an entry whose key is 5, highlighted in Figure 3-2), so linear probing next checks `table[6]`, which is also unavailable, and so the entry is placed in the next available bucket, `table[0]` (because open addressing wraps around when searching for the next available empty index position). Adding an entry with key 19 similarly causes a collision *with every nonempty bucket*, and the entry is finally placed in `table[2]`.

Using the existing `Hashtable` in Figure 3-2, I can `put` one additional entry into the table (since I need to leave one bucket empty). Where would I place an entry with a key of 44? The hash code for its key is 44 % 7 = 2, which is occupied; searching for the next available bucket places this entry in `table[3]`. Since `get()` and `put()` use the same search strategy, this entry will eventually be found later when invoking `get(44)`.

A *chain* of entries for hash code `hc` in an open addressing `Hashtable` is a sequence of consecutive `Entry` objects in `table`. This sequence starts at a given `table[hc]` and extends to the right (wrapping around to the first index in `table` as needed), up to *but not including* the next available unused `table` index position. In Figure 3-2, the chain for hash code 5 has a length of 5 even though there are only three entries whose key has that hash code. Also, there are no keys with a hash code of 2, but the chain for hash code 2 has a length of 1 because of an earlier collision. The maximum length of any chain is M – 1 because one bucket is left empty.

Listing 3-4 shows the modifications to `Hashtable` to support open addressing; the `Entry` class is unchanged. `Hashtable` keeps a running count, N, of the number of entries it stores so it can ensure there is at least one unused bucket (though it does not need to keep track of where that empty bucket is!). This is critically important to ensure the `while` loop in the `get()` and `put()` functions will eventually terminate.

Listing 3-4. Open addressing implementation of Hashtable

```
class Hashtable:
  def __init__(self, M=10):
    self.table = [None] * M
    self.M = M
    self.N = 0

  def get(self, k):
    hc = hash(k) % self.M                ❶
    while self.table[hc]:
      if self.table[hc].key == k:        ❷
        return self.table[hc].value
      hc = (hc + 1) % self.M             ❸
    return None                          ❹

  def put(self, k, v):
    hc = hash(k) % self.M                ❶
    while self.table[hc]:
      if self.table[hc].key == k:        ❺
        self.table[hc].value = v
        return
      hc = (hc + 1) % self.M             ❸

    if self.N >= self.M - 1:             ❻
      raise RuntimeError ('Table is Full.')

    self.table[hc] = Entry(k, v)         ❼
    self.N += 1
```

❶ Start at the first bucket where an entry whose key is k could be.

❷ If found, return the value associated with k.

❸ Otherwise advance to the next bucket, wrapping around to 0 as needed.

❹ If table[hc] is empty, you know k is not in table.

❺ If found, update the value associated with the Entry for k.

❻ Once you've established that k is not in table, throw RuntimeError if hc is the last empty bucket.

❼ Create new Entry in table[hc] and update N, the count of keys.

Given this new mechanism, what is the performance of get() and put()? Is the time to perform these operations independent of N, the number of keys in the Hashtable? No!

The Liberating Linked List

The most commonly used dynamic data structure is a *linked list*. Instead of allocating a contiguous block of memory, as needed for an array, a linked list stores data in memory fragments called *nodes* that are linked together so the programmer can start at a designated *first* node, and follow *links* (also called *references*) to the *next node* in the list when searching for a desired target.

The following linked list consists of three nodes, each of which stores a different value. Each node has a next reference (i.e., an arrow) to the next node in the list. The next reference in the final node in a list is None.

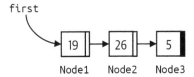

The size of a linked list is the number of nodes you encounter starting at *first* and traversing to each subsequent node using *next* until None is reached. The nodes are dynamically allocated memory. For object-oriented languages, nodes are objects. Both Java and Python rely on automatic memory to reclaim this memory when nodes are removed from a linked list, though other programming languages may require the programmer to manually release the memory when nodes are removed from a linked list.

Prepending a value
> You can prepend a value to the head of this list in constant time by creating a new node, Node0, to store this value and set Node0's *next* reference to Node1; then set *first* to point to Node0.

Appending a value
> If you maintain a reference to the *last* node in the list, you can append a value to the end of this list in constant time by creating a new node, Node4, to store this value; then set the *next* reference of *last* to point to Node4, and set *last* to Node4.

Inserting a value
> You can insert a value after an existing node, p, by creating a new node, q, to store this value and setting q.next to p.next, and then p.next = q.

Deleting a value
> You can delete a value from a linked list by locating its node in the linked list and adjusting *next* references accordingly, paying special attention if this value is the first one in the list.

Consider what happens in a *worst case* scenario. To an empty `Hashtable`, add an entry, and assume it is placed in `table[0]`. What if each of the next N – 1 requests to add an entry collides with this existing key in the `table` storage? What is the total number of buckets searched? For the first request, it is 1; for the second request, 2, and for the third request, 3. As you can see, there will be k buckets inspected for request number k. The total number, then, is the sum of 1 + 2 + ... + (N – 1), which conveniently equals N × (N – 1)/2. The *average* computation is this quantity divided by N, which leaves (N – 1)/2.

 Based on what I presented in Chapter 2, I can classify (N – 1)/2 as O(N) as follows: First, this formula can be written as N/2 – ½. As the problem instance size increases, the dominant term is N/2. In the *worst case*, the average number of buckets searched is directly proportional to N (in this case, half of N).

I have just demonstrated that in the *worst case*, the average number of buckets inspected is O(N). I use algorithmic analysis here to estimate the count of buckets (not performance time or number of operations) since the runtime performance for `get()` and `put()` is directly related to the number of buckets inspected.

This is a stalemate. You can increase the allocated `table` size, M, to be much greater than the number of keys to be inserted, N, and reduce the number of collisions and the overall time to execute. But if you don't plan correctly—that is, if N becomes closer and closer to M—you could quickly have horrible performance: even worse, the `table` can still fill up, preventing more than M – 1 entries from being stored. Table 3-2 compares the performance of inserting N entries into a `Hashtable` of size M that uses open addressing to resolve conflicts. Observe the following:

- For a small value of N, say 32, the average cost is nearly the same (read the values from left to right in that row) regardless of the size, M, of the `Hashtable`, because M is so much larger than N.

- For any `Hashtable` of size M, the average time to insert N keys consistently increases as N increases (read the numbers from top to bottom in any column).

- If you look "diagonally southeast" in the table, you will see that the timing results are more or less the same. In other words, if you want the average performance of inserting 2 × N keys to be the same as inserting N keys, then you need to double the initial size of the `Hashtable`.

Table 3-2. Average performance to insert N keys into a Hashtable *of size M (in milliseconds)*

	8,192	16,384	32,768	65,536	131,072	262,144	524,288	1,048,576
32	0.048	0.036	0.051	0.027	0.033	0.034	0.032	0.032
64	0.070	0.066	0.054	0.047	0.036	0.035	0.033	0.032
128	0.120	0.092	0.065	0.055	0.040	0.036	0.034	0.032
256	0.221	0.119	0.086	0.053	0.043	0.038	0.035	0.033
512	0.414	0.230	0.130	0.079	0.059	0.044	0.039	0.035
1,024	0.841	0.432	0.233	0.132	0.083	0.058	0.045	0.039
2,048	1.775	0.871	0.444	0.236	0.155	0.089	0.060	0.047
4,096	3.966	1.824	0.887	0.457	0.255	0.144	0.090	0.060
8,192	–	4.266	2.182	0.944	0.517	0.276	0.152	0.095
16,384	–	–	3.864	1.812	0.908	0.484	0.270	0.148

This working implementation of the *symbol table* data type only becomes efficient when the available storage is noticeably larger than the number of keys to be inserted. If you misjudge the total number of keys in the symbol table, then performance will be surprisingly inefficient, sometimes 100 times slower. Worse, it is not really useful since I have not yet added the capability to remove a key from the symbol table. To overcome these limitations, I need to introduce the *linked list* data structure.

Separate Chaining with Linked Lists

I now modify Hashtable to store an array of linked lists in a technique known as *separate chaining*. Where *linear probing* looked for an empty bucket in which to place an entry, separate chaining stores an array of linked lists where each linked list contains entries whose keys compute to the same hash code. These linked lists are formed from LinkedEntry nodes, shown in Listing 3-5.

Listing 3-5. LinkedEntry *node structure to support linked list of (key, value) pairs*

```
class LinkedEntry:
  def __init__(self, k, v, rest=None):
    self.key = k
    self.value = v
    self.next = rest        ❶
```

❶ rest is an optional argument, which allows the newly created node to link directly to an existing list pointed to by rest.

More accurately, `table[idx]` refers to the first `LinkedEntry` node in a linked list of nodes whose keys share the same hash code value of `idx`. For convenience, I still refer to the M buckets, which can be empty (in which case `table[idx]` = None) or contain the first `LinkedEntry` node in a linked list.

> The *chain* concept, introduced earlier for open addressing, is more clearly visible with linked lists: the length of the linked list is the length of the chain.

I continue to use `hash(key)` % M to compute the hash code for an entry to be inserted. All entries with the same hash code exist within the same linked list. In Figure 3-3, the `table` array has seven buckets, and therefore seven potential linked lists; once again, only the keys are shown.

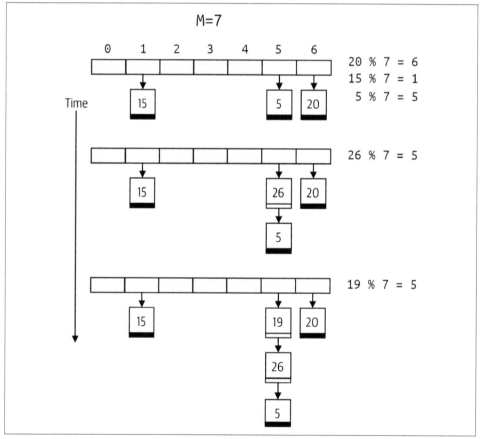

Figure 3-3. Structure of `Hashtable` linked list storage after adding five (key, value) pairs

Figure 3-3 adds the (key, value) pairs in the same order as Figure 3-2: the first three entries with keys 20, 15, and 5 create three linked lists, one for each of the buckets associated with the hash code for these keys. Buckets shaded in gray are empty. Adding an entry whose key is 26 causes a collision in table[5], so a new entry is *prepended* to the beginning of this linked list, resulting in a linked list with two entries. When adding the final entry whose key is 19, it is also prepended to the linked list in table[5], resulting in a linked list with three entries.

Pay attention to how the newest entry added to a bucket becomes the first entry in that bucket's linked list. Once put() determines that the entry's key does not appear in the linked list, it is efficient to just prepend a new LinkedEntry node at the beginning of the linked list. When the next reference is None, that entry is the last one in the linked list. Listing 3-6 shows the modifications to Hashtable.

Listing 3-6. Separate chaining implementation of Hashtable

```
class Hashtable:
  def __init__(self, M=10):
    self.table = [None] * M
    self.M = M
    self.N = 0

  def get(self, k):
    hc = hash(k) % self.M            ❶
    entry = self.table[hc]           ❷
    while entry:
      if entry.key == k:             ❸
        return entry.value
      entry = entry.next
    return None

  def put(self, k, v):
    hc = hash(k) % self.M            ❶
    entry = self.table[hc]           ❷
    while entry:
      if entry.key == k:             ❸
        entry.value = v              ❹
        return
      entry = entry.next

    self.table[hc] = LinkedEntry(k, v, self.table[hc])   ❺
    self.N += 1
```

❶ Compute index position, hc, of linked list for hash code that could contain k.

❷ Start with first node in the linked list.

❸ Traverse next reference until you find entry whose key matches k.

❹ Overwrite value associated with k.

❺ Prepend new node for (k, v) in table[hc] and increment count, N.

The get() and put() functions have nearly identical structures, with a while loop that visits each LinkedEntry node in the linked list. Starting with the first LinkedEntry in table[hc], the while loop visits each node exactly once until entry is None, which means all nodes were visited. As long as entry is not None, the entry.key attribute is inspected to see if there is an exact match with the k parameter. For get(), the associated entry.value is returned, while for put(), this value is overwritten with v. In both cases, in the *worst case*, the while loop will terminate once all entries in the linked list have been seen. When get() exhausts the linked list, it returns None since no entry was found; put() *prepends* a new entry (adds it at the beginning of the list), since it didn't exist before.

 put(k, v) only adds a new entry to a linked list after checking that none of the existing entries in the linked list have k as its key.

To evaluate the performance of this linked list structure, I need to count both the number of times a bucket is accessed as well as *the number of times an entry node is inspected*. It turns out that the performance of this linked list implementation is nearly identical to the open addressing implementation: the primary improvement is that there is no limit to the number of entries you can store in a linked list implementation. However, as I will discuss later in this chapter, if the number of entries, N, far exceeds the number of buckets, M, then the performance will degrade significantly. You also must consider that the memory requirement for storing the next references in the linked lists is twice as much as for open addressing.

Removing an Entry from a Linked List

Linked lists are versatile and dynamic data structures that can be extended, shortened, and spliced together efficiently because they do not rely on a fixed-size block of sequential memory. You can't remove an index position from an array, but you can remove a node from a linked list.

Let's break it down into two cases using a linked list containing three entries whose key values are shown in Figure 3-4. Let's say you want to delete the entry whose key is 19. This is the first node in the linked list. To remove this entry, simply set the value of `first` to be `first.next`. The modified linked list now has two entries, starting with 26.

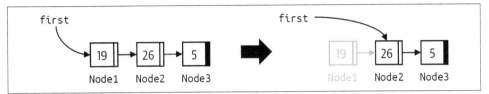

Figure 3-4. Removing the first node in a linked list

To delete any other `entry` (say, the one whose key is 26), look through the list, as shown in Figure 3-5. Stop when you find an entry with that key value, *but keep a reference to the prior node*, `prev`, *during your search*. Now set the value of `prev.next` to be `entry.next`. This cuts out this middle node and results in a linked list containing one fewer node. Note that this will also handle the case when `entry` is the final entry in the linked list, since then `prev.next` is set to `None`.

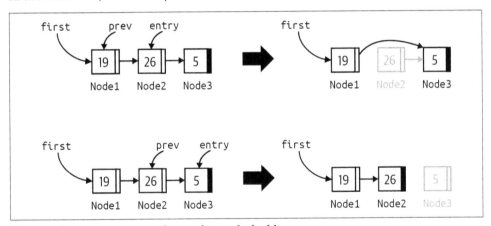

Figure 3-5. Removing any other node in a linked list

In both cases, the memory for the removed node is reclaimed. Listing 3-7 contains the `remove(k)` method in the `Hashtable` linked list implementation that removes the entry with the given key, should it exist.

Listing 3-7. remove() function for separate chaining implementation of Hashtable

```
def remove(self, k):
  hc = hash(k) % self.M
  entry = self.table[hc]              ❶
  prev = None
  while entry:                        ❷
    if entry.key == k:                ❸
      if prev:
        prev.next = entry.next        ❹
      else:
        self.table[hc] = entry.next   ❺

      self.N -= 1                      ❻
      return entry.value

    prev, entry = entry, entry.next    ❼

  return None
```

❶ self.table[hc] refers to the first entry in the linked list associated with hash code hc.

❷ Continue iterating as long as there are entries in this linked list.

❸ Locate entry to remove by comparing target k with the key field for entry.

❹ When found, if there is a prev reference, link around entry, which removes it from the list.

❺ If there is no prev reference, then entry is first. Set the linked list at self.table[hc] to point to the second node in the linked list.

❻ Decrement count of entries, N. It is common to return the value associated with the entry being removed.

❼ If key is not found, continue iteration by setting prev to entry and advancing entry to the next entry.

By convention, the remove(k) function returns the value that had been associated with k or None if k was not present.

Evaluation

I now have two different structures that provide a symbol table data type to store (key, value) pairs. The linked list implementation has the added benefit of allowing (key, value) pairs to be removed, so you must choose that structure if you need this functionality. If you only add entries to a symbol table, however, you still need to evaluate the efficiency of these two approaches.

Let's first evaluate the storage requirements to store N (key, value) pairs. Both approaches create an array of size M to hold entries. However, in the linked list approach, N can grow to be as large as necessary; for open addressing, N must be strictly smaller than M, so you must be sure to choose a large enough M in advance. The size of the memory requirements for `table` is directly proportional to M.

Ultimately, there will be N entries inserted into the symbol table. Each `Entry` in open addressing stores only the (key, value) pair, while the `LinkedEntry` for the linked list approach stores an additional `next` reference for each entry. Since each reference is a fixed memory size, the extra storage is directly proportional to N.

- Open addressing requires storage proportional to both M and N, but since N < M, I can simply say that the storage is O(M).
- Separate chaining requires storage proportional to both M and N, but since there are no restrictions on N, the storage is O(M + N).

To evaluate runtime performance, the key operation to count is the number of times an entry is inspected. Let's start with the *worst case*, which occurs when the computed hash code for all keys is the same. In the linked list implementation, the `table` array contains M − 1 unused index positions, and a single linked list contains all N (key, value) pairs. The key to be searched might be the final entry in the linked list, so the `get()` performance in the *worst case* would be directly proportional to N. The situation in open addressing is similar: there would be N consecutive entries within the M-sized array, and the one you are looking for is the last one. I can safely say that *regardless of implementation choice*, in the *worst case*, `get()` is O(N).

This might seem like a deal-breaker, but it turns out that the mathematical hash functions do a really good job of distributing the hash code values for keys; as M increases, the probability of collisions decreases. Table 3-3 describes a head-to-head comparison of these two approaches by inserting N = 321,129 words from an English dictionary into a hashtable whose size, M, varies from N/2 to 2 × N. It also includes results for M = 20 × N (first row) and smaller values of M (the last five rows).

Table 3-3 shows two pieces of information for each (M, N) pair:

- The average length of each nonempty chain in the `Hashtable`. This concept is applicable whether open addressing or linked lists are used.

- The size of the maximum chain in the `Hashtable`. If the open addressing `Hashtable` becomes too congested—or linked lists become excessively long for certain hash codes—the runtime performance suffers.

Table 3-3. Average performance when inserting N = 321,129 keys into a Hashtable of size M as M decreases in size

M	Linked list		Open addressing	
	Avg. chain len	Max chain len	Avg. chain len	Max chain len
6,422,580	1.0	4	1.1	6
...	...	...	...	...
642,258	1.3	6	3.0	44
610,145	1.3	7	3.3	46
579,637	1.3	7	3.6	52
550,655	1.3	7	4.1	85
523,122	1.3	7	4.7	81
496,965	1.4	7	5.4	104
472,116	1.4	7	6.4	102
448,510	1.4	7	7.8	146
426,084	1.4	7	10.1	174
404,779	1.4	7	14.5	207
384,540	1.5	7	22.2	379
365,313	1.5	9	40.2	761
347,047	1.5	9	100.4	1429
329,694	1.6	8	611.1	6735
313,209	1.6	9	Fail	
...	...	...	...	
187,925	2.1	9	Fail	
112,755	3.0	13	Fail	
67,653	4.8	16	Fail	
40,591	7.9	22	Fail	
24,354	13.2	29	Fail	

The values in the table all increase as the size of M decreases; this occurs because smaller `Hashtable` sizes lead to more collisions, producing longer chains. As you can see, however, open addressing degrades much more quickly, especially when you consider that there are some hash codes whose chain extends to hundreds of entries to be inspected. Worse, once M is smaller than N, it becomes impossible to use open addressing (shown in the table as `Fail`). In contrast, the statistics for the linked list implementation appear to be almost immune from this situation. If the size, M, of a hashtable is much larger than N—for example, twice as large—then the average chain length is very close to 1, and even the maximum chain length is quite small. However, M *has to be decided in advance*, and if you use open addressing, you will run out of room once N = M – 1.

Things are much more promising with separate chaining. As you can see in Table 3-3, even when N is more than ten times the size of the hashtable, M, linked lists can grow to accommodate all of the entries, and the performance doesn't suffer nearly as much as open addressing does when N became closer and closer to M. This is evident from the maximum chain length of the linked lists shown in Table 3-3.

These numbers provide the information to develop a strategy to ensure the efficiency of a hashtable whose initial size is M. The performance of a hashtable can be measured by how "full" it is—which can be determined by dividing N by M. Mathematicians have even defined the term *alpha* to represent the ratio N/M; computer scientists refer to *alpha* as the *load factor* of a hashtable.

- For separate chaining, *alpha* represents *the average number of keys in each linked list* and can be larger than 1, limited only by the available memory.
- For open addressing, *alpha* is the percentage of buckets that are occupied; its highest value is (M – 1)/M, so *it must be smaller than 1*.

Years of research have identified that hashtables become increasingly inefficient once the load factor is higher than 0.75—in other words, once an open addressing hashtable is ¾ full.[7] For separate chaining hashtables, the concept still applies, even though they do not "fill up" in the same way.

Figure 3-6 plots the average chain size (shown in diamonds using the left Y-axis) and maximum chain size (shown in squares using the right Y-axis) after inserting N = 321,129 words into a `Hashtable` of size M (found on the X-axis). This graph effectively shows how you can compute a suitable M value to ensure a desired average (or maximum) chain size if you know the number of keys, N, to be inserted.

---

7 The Python `dict` type uses ⅔ as the threshold.

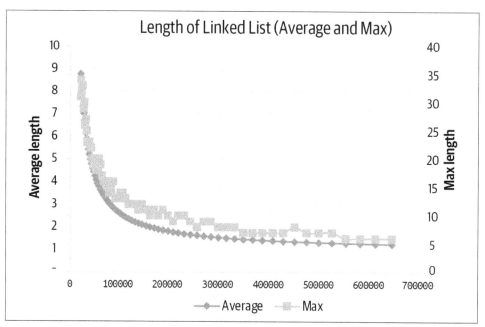

Figure 3-6. *For fixed number of elements N, average and maximum chain length follow predictable paths*

If the hashtable could only grow larger—that is, increase its M value—then the load factor would reduce and the hashtable would be efficient again. I next show how to accomplish this, with a little bit of effort.

Growing Hashtables

The DynamicHashtable in Listing 3-8 uses a load_factor of 0.75 to set a threshold target.

Listing 3-8. Determine load_factor and threshold when creating DynamicHashtable

```
class DynamicHashtable:
  def __init__(self, M=10):
    self.table = [None] * M
    self.M = M
    self.N = 0

    self.load_factor = 0.75
    self.threshold = min(M * self.load_factor, M-1) ❶
```

❶ Ensure for M ≤ 3 that threshold is no greater than M − 1.

What if I simply doubled the size of the storage array, using a resize strategy known as *geometric resizing*? More precisely, double the array size and add 1 when resizing.[8] Once the number of (key, value) pairs is larger than or equal to threshold, the table storage array needs to grow in size to remain efficient. The revised put() method for separate chaining is shown in Listing 3-9.

Listing 3-9. Revised put() method initiates resize()

```
def put(self, k, v):
  hc = hash(k) % self.M
  entry = self.table[hc]
  while entry:
    if entry.key == k:
      entry.value = v
      return
    entry = entry.next

  self.table[hc] = LinkedEntry(k, v, self.table[hc])  ❶
  self.N += 1

  if self.N >= self.threshold:                         ❷
    self.resize(2*self.M + 1)                          ❸
```

❶ Prepend new entry to the table[hc] linked list chain.

❷ Check whether N meets or exceeds threshold for resizing.

❸ Resize storage array into new array that is twice original size plus one.

To increase storage in most programming languages, you need to allocate a new array and copy over all of the original entries from the first array into the second array, as shown in Figure 3-7. This figure shows the result after resizing the array storage for the array of linked lists as well as open addressing.

You should first observe that this copy operation will take time that is directly proportional to M—the greater the size of the hashtable storage array, the more elements need to be copied—so this operation can be classified as O(M). Simply copying the entries won't actually work, however, since some entries—such as the ones with 19 and 26 as keys—*will no longer be findable.*

---

8 It is commonly observed that Hashtables whose number of buckets is a prime number work really well (see challenge exercise at end of this chapter); here the size is odd, which can also be helpful.

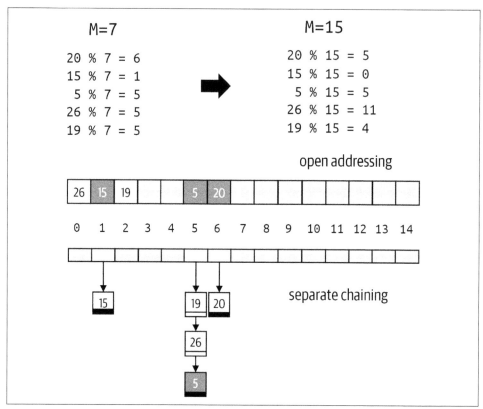

Figure 3-7. Some entries can get "lost" if they are simply copied when M increases

If you went to look for key 19, its hash code would now be 19 % 15 = 4, and that bucket is empty in both structures, indicating that no entry with a key of 19 exists in the hashtable. In the former open addressing Hashtable with size M = 7, key 19 had been placed in bucket 2 because linear probing wrapped around the end of the old array (of size 7) to the beginning when it was inserted. Now that the array has 15 elements, the wraparound doesn't happen, so this key can no longer be found.

By pure coincidence, some entries are still findable, and these are highlighted in Figure 3-7. In open addressing, the entry with key 20 is not in table[5], but linear probing will find it in table[6]; similarly, the entry with key 15 is not in table[0], but linear probing will find it in table[1]. The entry with key 5 is findable in both open addressing and separate chaining because its hash code remains the same.

What can I do to avoid losing keys? The proper solution, shown in Listing 3-10, is to create a new temporary Hashtable with twice the original storage (technically, 2M + 1) and *rehash* all entries into the new Hashtable. In other words, for each of the (k, v) entries in the original Hashtable, call put(k,v) on the new temporary Hashta ble. Doing so will guarantee these entries will remain findable. Then, with a nifty programming trick, the underlying storage array from temp is stolen and used as is, whether for separate chaining or for open addressing.

Listing 3-10. Resize method to dynamically grow hashtable storage for separate chaining

```
def resize(self, new_size):
  temp = DynamicHashtable(new_size)          ❶
  for n in self.table:
    while n:
      temp.put(n.key, n.value)               ❷
      n = n.next
  self.table = temp.table                    ❸
  self.M = temp.M                            ❹
  self.threshold = self.load_factor * self.M
```

❶ Construct temporary Hashtable of desired new size.

❷ For each node in the linked list for a given bucket, take all nodes and rehash each entry into temp.

❸ Grab the storage array from temp and use for our own.

❹ Be sure to update our M and threshold values.

This code is nearly identical for resizing with open addressing. I now have a strategy for dynamically increasing the size of a Hashtable. Figure 3-8 contains the proper resized hashtables for the open addressing example in Figure 3-2 and the separate chaining example in Figure 3-3.

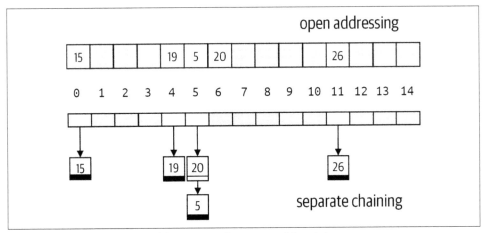

Figure 3-8. Resulting hashtable storage after successful resizing

How well does the resizing logic perform? Let's conduct an experiment running 25 repeated trials using different M values to measure:

Build time
> The time it takes to add N = 321,129 keys into a `Hashtable` with initial size M that doubles in size when `threshold` is exceeded.

Access time
> The time to locate all N of these words once all keys are inserted.

For separate chaining and open addressing hashtables, Table 3-4 compares the build times and access times for dynamically resizable hashtables starting with different initial size, M, ranging from 625 to 640,000. This table also shows the performance of nongrowing hashtables with an initial size of M = 428,172. This will result in a fair comparison, since N = 321,129, or 428,172 × 0.75.

Table 3-4. Comparing growing tables against fixed-size construction (time in ms)

M	Linked list Build time	Linked list Access time	Open addressing Build time	Open addressing Access time
625	0.997	0.132	1.183	0.127
1,250	1.007	0.128	1.181	0.126
2,500	0.999	0.129	1.185	0.133
5,000	0.999	0.128	1.181	0.126
10,000	1.001	0.128	1.184	0.126
20,000	0.993	0.128	1.174	0.126
40,000	0.980	0.128	1.149	0.125
80,000	0.951	0.130	1.140	0.127
160,000	0.903	0.136	1.043	0.126
320,000	0.730	0.132	0.846	0.127
640,000	0.387	0.130	0.404	0.127
...	...	...	...	...
Fixed	0.380	0.130	0.535	0.131

The last row presents the ideal case, where the load factor of the `Hashtable` does not exceed 0.75. You should expect open addressing to require more build time, because collisions on one bucket inevitably affect other buckets, whereas separate chaining localizes all collisions within the same bucket.

The other remaining rows show that you don't have to magically predict the initial M to use, since the access time (to inspect all 321,129 keys) is essentially the same.

Analyzing the Performance of Dynamic Hashtables

In the *worst case*, both put() and get() for Hashtable is O(N). As I have explained before, if the hash code for each key computes to exactly the same bucket (for both separate chaining and open addressing), the time to complete each operation is directly proportional to N, the number of keys in the Hashtable.

However, under the commonly agreed-upon assumption of *simple uniform hashing*, hash functions will uniformly distribute keys into the buckets of the hashtable: each key has an equal probability of being placed in any bucket. From this assumption, mathematicians have proved that the average length of each chain is N/M. The take-away is that you can rely on the experts who have developed the Python hash() function.

Since you can always guarantee that N < M when hashtables can grow, I can say for certain that the average quantity N/M is a constant, O(1), and is *independent of* N.

 A search can either hit (the key is in the hashtable) or miss (it is not). In open addressing (assuming uniform key distribution), on a hit the average number of buckets to inspect is $(1 + 1/(1 - alpha))/2$. This works out to 2.5 when $alpha = 0.75$. For a search miss, the result is $(1 + 1 / (1 - alpha)^2)/2$. This works out to 8.5 under the same assumption.

I need to account for the extra costs of resizing the hashtable, working under the assumption that the threshold load factor is 0.75 and that the underlying storage doubles in size using geometric resizing. To start things off, let's assume M is 1,023 and that N is much larger than M: I'll use the 321,129-word English dictionary again. I need to count the number of times each key is inserted into a hashtable (including the temporary one in resize).

The first resize is triggered when the 768th key is added (since 768 is ≥ 767.25 = 0.75 × 1,023), which grows the hashtable to have M = 1,023 × 2 + 1, or 2,047. During this resize, 768 keys are rehashed and inserted. Note that immediately after the resize, the load factor is reduced in half to 768/2047, which is approximately 0.375.

When an additional 768 keys are inserted, all 1,536 keys are rehashed into a new hashtable of size M = 2,047 × 2 + 1, or 4,095. The hashtable is resized a third time when an additional 1,536 keys are inserted, which causes the existing 3,072 keys to be inserted into a hashtable of size M = 8,191.

To make sense of these numbers, Table 3-5 shows the resizing moments when the Nth word is inserted, together with the accumulated total number of times any key is inserted into the hashtable. During resize events, the final column shows that the average number of insertions (by dividing the total number of insertions by N) converges to 3. Even though resizing forces each key to be reinserted with a geometric resizing strategy, this table demonstrates you will never need more than three times as many insertions than without resizing.

Table 3-5. Words whose addition causes a resize event, with total number of insertions and average number of times a word was inserted

Word	M	N	# insert	average
absinths	1,023	768	1,536	2.00
accumulatively	2,047	1,536	3,840	2.50
addressful	4,095	3,072	8,448	2.75
aladinist	8,191	6,144	17,664	2.88
anthoid	16,383	12,288	36,096	2.94
basirhinal	32,767	24,576	72,960	2.97
cincinnatian	65,535	49,152	146,688	2.98
flabella	131,071	98,304	294,144	2.99
peps	262,143	196,608	589,056	3.00
...	...	...	...	...
zyzzyvas	524,287	321,129	713,577	2.22

The key observation is that geometric resizing ensures that resize events occur *significantly less frequently* as the size of the table grows. In Table 3-5, once the final word is inserted into the hashtable—which is not a resize event—the average has dropped to 2.22, and it will not need to be resized again until another 72,087 keys are added. As you can see, this is almost 100 times less frequently than when it started (when a resize was triggered after just 768 were added).

The result of this last analysis is that the average cost of inserting all 321,129 entries into a dynamically resizing hashtable is *no more than three times* what it would cost if the hashtable had somehow been large enough to store all N keys. As you have seen in Chapter 2, this is just a multiplicative constant, which will not change the performance classification of the average case for put(): it remains O(1) even with extra work due to resizing.

Perfect Hashing

If you know the collection of N keys in advance, then you can use a technique known as *perfect hashing* to construct an optimal hashtable where the hash code for each key is a unique index location. Perfect hashing *generates the Python code* containing the hash function to use. This is an unexpected result, and there are perfect hash generators for numerous programming languages.

If you install the third-party Python library perfect-hash, it can generate a per
fect_hash() function from an input file containing the desired keys.[9] Listing 3-11
contains the generated code using the words "a rose by any other name would smell
as sweet."

Listing 3-11. Perfect hashtable for ten words from Shakespeare

```
G = [0, 8, 1, 4, 7, 10, 2, 0, 9, 11, 1, 5]

S1 = [9, 4, 8, 6, 6]
S2 = [2, 10, 6, 3, 5]

def hash_f(key, T):
  return sum(T[i % 5] * ord(c) for i, c in enumerate(key)) % 12

def perfect_hash(key):
  return (G[hash_f(key, S1)] + G[hash_f(key, S2)]) % 12
```

The enumerate() built-in Python function improves how you iterate over lists when you also need positional information.

```
>>> for i,v in enumerate(['g', 't', 'h']):
        print(i,v)
0 g
1 t
2 h
```

enumerate() iterates over each value in a collection and addition-
ally returns the index position.

Recall from Figure 3-1 how I defined a list, day_array, and a supporting base26()
hash function that worked with twelve months? Perfect hashing pursues the same
approach in a more elegant way, processing the N strings to create the lists G, S1, and
S2 and a supporting hash_f() function.

To compute the index location for the string 'a', you need two intermediate results;
recall that ord('a') = 97:

- hash_f('a', S1) = sum([S1[0] × 97]) % 12. Since S1[0] = 9, this is the value (9
  × 97) % 12 = 873 % 12 = 9.

- hash_f('a', S2) = sum([S2[0] × 97]) % 12. Since S2[0] = 2, this is the value (2
  × 97) % 12 = 194 % 12 = 2.

---

9 Use the Python pip installer like this: pip install perfect-hash.

The value returned by perfect_hash('a') is (G[9] + G[2]) % 12 = (11 + 1) % 12 = 0. This means that the hash code for the string 'a' is 0. Repeat this computation[10] for the string 'by' and you will find that:

- hash_f('by', S1) = $(9 \times 98 + 4 \times 121)$ % 12 = 1,366 % 12 = 10.
- hash_f('by', S2) = $(2 \times 98 + 10 \times 121)$ % 12 = 1,406 % 12 = 2.
- perfect_hash('by') is (G[10] + G[2]) % 12 = (1 + 1) % 12 = 2.

To summarize, the key 'a' hashes to index location 0, and the key 'by' hashes to index location 2. In fact, each of the words in "a rose by any other name would smell as sweet" is hashed to a different computed index location. It is sometimes truly amazing what mathematics can do!

Listing 3-12 contains the perfect_hash() function for the 321,129 words in the sample dictionary. This function computes 0 for the first English word, 'a', and 321,128 for the last English word, 'zyzzyvas'. It is supported by a large list, G, containing 667,596 values (not shown, obviously!) and two intermediate lists, S1 and S2.

For the string 'by' in this larger perfect hashtable, you can confirm the following:

- hash_f('by', S1) = $(394{,}429 \times 98 + 442{,}829 \times 121)$ % 667,596 = 92,236,351 % 667,596 = 108,103
- hash_f('by', S2) = $(14{,}818 \times 98 + 548{,}808 \times 121)$ % 667,596 = 67,857,932 % 667,596 = 430,736
- perfect_hash('by') = (G[108,103] + G[430,736]) % 667,596 = (561,026 + 144,348) % 667,596 = 37,778

Listing 3-12. Partial listing of perfect hash function for English dictionary

```
S1 = [394429, 442829, 389061, 136566, 537577, 558931, 481136,
      337378, 395026, 636436, 558331, 393947, 181052, 350962, 657918,
      442256, 656403, 479021, 184627, 409466, 359189, 548390, 241079, 140332]
S2 = [14818, 548808, 42870, 468503, 590735, 445137, 97305,
      627438, 8414, 453622, 218266, 510448, 76449, 521137, 259248, 371823,
      577752, 34220, 325274, 162792, 528708, 545719, 333385, 14216]

def hash_f(key, T):
  return sum(T[i % 24] * ord(c) for i, c in enumerate(key)) % 667596

def perfect_hash(key):
  return (G[hash_f(key, S1)] + G[hash_f(key, S2)]) % 667596
```

---

10 ord('b') = 98, and ord('y') = 121.

The computation in `perfect_hash(key)` in Listing 3-12 produces a large sum that is reduced using % 667,596 to identify a unique location from the large G list. As long as key is a valid English word from the dictionary, `perfect_hash(key)` uniquely identifies an index from 0 to 321,128.

If you inadvertently attempt to hash a key that is not an English word, a collision will occur: the word `'watered'` and the nonword `'not-a-word'` both hash to the index location 313,794. This is not an issue for perfect hashing, since the programmer is responsible for ensuring that only valid keys are ever hashed.

Iterate Over (key, value) Pairs

Hashtables are designed for efficient `get(k)` and `put(k,v)` operations. It might also be useful to retrieve all entries in a hashtable, whether it uses open addressing or separate chaining.

Python generators are one of the best features of the language. Most programming languages force programmers to return the values in a collection using extra storage. In Chapter 2, I explained how `range(0, 1000)` and `range(0, 100000)` both use the same amount of memory while returning all integers in the respective ranges; generators make this possible.

The following generator function produces the integers in the range from 0 to n that do not include the given `digit`:

```
def avoid_digit(n, digit):
  sd = str(digit)
  for i in range(n):
    if not sd in str(i):
      yield i
```

To give an object this same capability in Python, a class can provide an `__iter__()` method that allows the caller to use the `for v in object` idiom.

The two `__iter__()` implementations in Listing 3-13 are designed for separate chaining and open addressing hashtables.

Listing 3-13. Iterate over all entries in a hashtable using Python generator function

```
# Iterator for Open Addressing Hashtable
def __iter__(self):
  for entry in self.table:
    if entry:                           ❶
      yield (entry.key, entry.value)    ❷

# Iterator for Separate Chaining Hashtable
def __iter__(self):
  for entry in self.table:
    while entry:                        ❸
      yield (entry.key, entry.value)    ❷
      entry = entry.next                ❹
```

❶ Skip over table entries that are None.

❷ Generate a tuple containing the key and `value` using Python `yield`.

❸ As long as there is a linked list at this bucket, yield a tuple for each node.

❹ Set `entry` to the next entry in the linked list, or None if none are left.

To demonstrate how these iterators work, construct two hashtables with M equal to 13—one using open addressing and one using separate chaining—and a third hashtable using perfect hashing. After inserting the words in the string "a rose by any other name would smell as sweet," Table 3-6 shows the words generated by the respective hashtables.

Table 3-6. Order of words returned by hashtable iterators

Open addressing	Separate chaining	Perfect hash
a	sweet	a
by	any	any
any	a	as
name	would	by
other	smell	name
would	other	other
smell	as	rose
as	name	smell
sweet	by	sweet
rose	rose	would

The words returned for the open addressing and separate chaining hashtables appear to be in a random order; of course, this isn't randomness but based solely on the way keys are hashed. If you execute the sample code that produces this table, your ordering for open addressing and separate chaining will likely be different, because the hash() code values for strings in Python 3 are unpredictable.

One nice feature of the perfect-hash library is that the index positions computed by perfect_hash(key) are based on the *order of the words used when generating the perfect hashing code*. Simply use a list of strings that are already in sorted order, and the entries will be stored in sorted order, and the iterator will yield the pairs in the same order.

Chapter 8 contains more details of the Python dict type. Instead of implementing a symbol table from scratch, as I have done in this chapter, you should always use dict because it is a built-in type and will be significantly more efficient than the code I provide.

Summary

In this chapter, I introduced several key concepts:

- The linked list data structure can store small collections of items efficiently, allowing for dynamic insertion and removal.
- Symbols tables use a storage array with M buckets to store entries. There needs to be a strategy to resolve collisions when two or more keys hash to the same bucket.
- Open addressing relies on distributing entries within an array to reduce the number of collisions, but this only works efficiently when the size of the storage array, M, is more than twice as large as the number of stored entries, N. See the challenge exercises for two approaches that support removing keys.
- Separate chaining uses linked lists to store entries whose keys hash to the same bucket. This approach makes it easier to support a remove operation for keys.
- Designing hash functions is hard—use the predefined ones designed by Python's designers.
- Geometric resizing ensures a symbol table remains efficient by decreasing the frequency of future resize events.
- Perfect hashing can be used to construct hash functions that avoid collisions by computing a unique bucket index location for a fixed number of keys; the hash function is often more computationally expensive than a default hash() function.

Challenge Exercises

1. Does *open addressing* improve if you use a different strategy for resolving collisions? Instead of using a *linear probe* with a delta of 1 (wrapping around), create a hashtable whose size is a power of 2, and use a probe sequence that explores additional index positions using deltas that are *triangle numbers*, that is, the integers 1, 3, 6, 10, 15, 21, and so on; you will still have to wrap around. The nth triangle number is the sum of the integers from 1 to n, represented by the formula $n \times (n + 1)/2$.

   Is the overall performance better than linear probing?

   Conduct an experiment where you fill a `Hashtable` of size 524,288 with the first 160,564 English words (for a utilization of 30%), and then measure the time it takes to search for all 321,129 words. Compare this performance against an open addressing `Hashtable` and the separate chaining variation.

2. Is it worth sorting the keys in the linked lists of a separate chaining hashtable? Construct a `Hashtable` variant that sorts the (key, value) pairs in each linked list in ascending order by key.

   Conduct an experiment where you measure the time it takes to construct the initial `Hashtable` of size 524,287 (a prime number) with the first 160,564 English words (for a utilization of 30%) *in reverse order*. As you will see, the *worst case* for this variation occurs when the keys are `put()` into the hashtable in increasing value. Compare this performance against an open addressing `Hashtable` and the regular separate chaining `Hashtable`.

   Now measure the time it takes to search the first 160,564 English words as keys. As you might expect, this is the *best case*, since all these words are in the hashtable. Compare this performance against the array-based `Hashtable` and the separate chaining variation. Next search for the remaining 160,565 words in the back half of the English dictionary. This provides a glimpse of the *worst case* since finding these words in a linked list will always require each linked list to be fully explored. Once again, compare this performance against the open addressing `Hashtable` and the regular unordered linked list variation.

   How sensitive are these results to the chosen initial size, 524,287? For example, compare with 214,129 (a utilization of 75%) and 999,983 (a utilization of 16%).

3. To see the dangers in predictable hash codes, consider the `ValueBadHash` code in Listing 3-14. The objects for this Python class hash to just four different values (0 to 3). This class overrides the default behavior of `hash()` as well as `__eq__()`, so these objects can be used as a key when invoking `put(key, v)`.

Listing 3-14. ValueBadHash has a horrible hash() function

```
class ValueBadHash:
  def __init__(self, v):
    self.v = v

  def __hash__(self):
    return hash(self.v) % 4

  def __eq__(self, other):
    return (self.__class__ == other.__class__ and self.v == other.v)
```

Construct a separate chaining `Hashtable(50,000)` and invoke `put(ValueBad Hash(w), 1)` for the first 20,000 English words in the dictionary. Next, create a regular separate chaining `Hashtable`, and invoke `put(w, 1)` for these same words. Generate statistics on:

- The average chain length of a bucket

- The maximum chain length of any bucket in `Hashtable`

Be prepared for the code execution to take a long time! Explain these statistics.

4. Having a prime number for the number of buckets, M, is helpful in practice because *every key that shares a common factor with M will be hashed to a bucket that is a multiple of this factor*. For example, if M = 632 = 8 × 79 and the entry to be inserted has a key of 2,133 = 27 × 79, the hash code is 2,133 % 632 = 237 and 237 = 79 × 3. The problem is that the performance of a hashtable assumes uniform distribution of keys, which is violated when some keys are predisposed to be placed in certain buckets.

   To demonstrate the impact of M, form a key from the `base26` representation of each of the words in the 321,129-word English dictionary. In the range from 428,880 to 428,980 containing 101 potential values for M, construct fixed-size hashtables (using both open addressing and separate chaining) and produce a table showing the average and maximum chain length size. Are there any values of M in this range that are particularly bad? Can you find out what these values *all have in common*? Armed with this knowledge, scan the next 10,000 higher values of M (up to 438,980) to try to find one whose maximum chain length is almost ten times as bad.

5. When using an open addressing hashtable, there is no easy way to support `remove(key)` because removing an entry from a bucket may break an existing chain created using linear probing. Imagine you have a `Hashtable` with M = 5 and you hashed entries with key values of 0, 5, and then 10. The resulting `table` array would contain the following keys [0, 5, 10, None, None], since each collision would be resolved using linear probing. A flawed `remove(5)` operation would simply remove the entry with key 5 from `table`, resulting in an array storing [0,

None, 10, None, None]. However, the entry with key 10 is no longer findable because the chain at index location 0 was broken.

One strategy is to add a Boolean field to the `Entry` that records whether the entry has been deleted or not. You have to modify `get()` and `put()` accordingly. In addition, the resize event needs to be adjusted as well, since entries marked as deleted will not need to be inserted into the new hashtable. Don't forget to update the `__iter__()` method to skip deleted entries.

Consider adding logic to initiate a shrink event when more than half of the entries in the hashtable are marked for deletion. Run some trials to compare the performance of separate chaining vs. open addressing now that `remove()` is available for open addressing.

6. Resizing a hashtable incurs a steep penalty since all N values need to be rehashed into the new structure. In *incremental resizing*, instead, the resize event allocates a new array, `newtable` (with size 2M + 1), but the original hashtable remains. A `get(key)` request first searches for key in `newtable` before searching through the original `table`. A `put(key,value)` request inserts a new entry into `newtable`: after each insertion, `delta` elements from the old `table` are rehashed and inserted into `newtable`. Once all elements are removed from the original `table`, it can be deleted.

   Implement this approach with a separate chaining hashtable, and experiment with different values of `delta`. In particular, what is the smallest value of `delta` that will ensure the old table is completely emptied before the next resize event? Is `delta` a constant, or is it based on M or even N?

   This approach will reduce the total cost incurred at any time by `put()`; perform empirical measurements using an example similar to Table 3-5, but now evaluating the runtime performance cost of the most expensive operation.

7. Should the number of (key, value) pairs, N, in a hashtable become less than ¼ of M, the `table` storage array could shrink to reduce unneeded storage space. This shrink event can be triggered by `remove()`.

   Modify either the separate chaining or open addressing hashtable to realize this capability, and run some empirical trials to determine if it is worthwhile.

8. Use a symbol table to find the element in a list that is duplicated the most number of times. In the case of a tie, any value that is repeated "the most number of times" can be returned.

   `most_duplicated([1,2,3,4])` can return 1, 2, 3, or 4, while `most_duplicated([1,2,1,3])` must return 1.

9. An open addressing hashtable can remove entries by rehashing all remaining (key, value) pairs in the chain *after the one that is to be removed*. Add this capabil-

ity to an open addressing hashtable, and run some trials to compare the performance of separate chaining versus open addressing with this revised remove capability.

## CHAPTER 4

Heaping It On

In this chapter, you will learn:

- The *queue* and *priority queue* data types.
- The *binary heap* data structure, invented in 1964, which can be stored in an array.
- That in a *max binary heap*, an entry with a *larger numeric priority* is considered to have higher priority than an entry with a smaller numeric priority. In a *min binary heap*, entries with smaller numeric priority have higher priority.
- How to enqueue a (value, priority) entry to a binary heap in O(log N), where N is the number of entries in the heap.
- How to find the value with highest priority in a binary heap in O(1).
- How to remove the value with highest priority from a binary heap in O(log N).

Instead of just storing a collection of values, what if you stored a collection of entries, where each entry has a *value* and an associated *priority* represented as a number? Given two entries, the one whose priority is higher is more important than the other. The challenge this time is to make it possible to insert new (value, priority) entries into a collection and be able to *remove and return the value for the entry with highest priority from the collection.*

This behavior defines a *priority queue*—a data type that efficiently supports enqueue(value, priority) and dequeue(), which removes the value with highest priority. The priority queue is different from the symbol table discussed in the previous chapter because *you do not need to know the priority in advance* when requesting to remove the value with highest priority.

When a busy nightclub becomes too crowded, a line of people forms outside, as shown in Figure 4-1. As more people try to enter the club, they have to wait at the end of the line. The first person to enter the nightclub from the line is the one who has waited the longest. This behavior represents the essential *queue* abstract data type, with an enqueue(value) operation that adds value to become the newest value at the end of the queue, and dequeue() removes the oldest value remaining in the queue. Another way to describe this experience is "First in, first out" (FIFO), which is short-hand for "First [one] in [line is the] first [one taken] out [of line]."

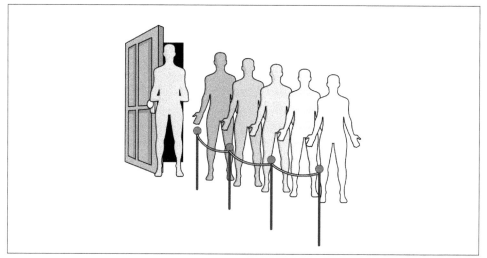

Figure 4-1. Waiting in a queue at a nightclub

In the previous chapter, I described the linked list data structure, which I will use again with a Node that stores a value for the queue:

```
class Node:
  def __init__(self, val):
    self.value = val
    self.next = None
```

Using this structure, the Queue implementation in Listing 4-1 has an enqueue() operation to add a value to the end of a linked list. Figure 4-2 shows the result of enqueuing (in this order) "Joe," "Jane," and "Jim" to a nightclub queue.

"Joe" will be the first patron dequeued from the line, resulting in a queue with two patrons, where "Jane" remains the first one in line.

In Queue, the enqueue() and dequeue() operations perform in constant time, independent of the total number of values in the queue.

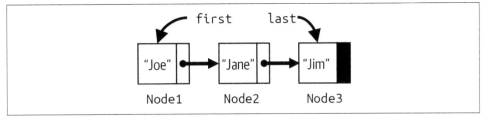

Figure 4-2. Modeling a nightclub queue with three nodes

Listing 4-1. Linked list implementation of Queue data type

```
class Queue:
  def __init__(self):
    self.first = None                           ❶
    self.last = None

  def is_empty(self):
    return self.first is None                   ❷

  def enqueue(self, val):
    if self.first is None:                      ❸
      self.first = self.last = Node(val)
    else:
      self.last.next = Node(val)                ❹
      self.last = self.last.next

  def dequeue(self):
    if self.is_empty():
      raise RuntimeError('Queue is empty')

    val = self.first.value                      ❺
    self.first = self.first.next                ❻
    return val
```

❶ Initially, first and last are None.

❷ A Queue is empty if first is None.

❸ If Queue is empty, set first and last to newly created Node.

❹ If Queue is nonempty, add after last, and adjust last to point to newly created Node.

❺ first refers to the Node containing value to be returned.

❻ Set first to be the second Node in the list, should it exist.

Let's change the situation: the nightclub decides to allow patrons to spend any amount of money to buy a special pass that records the total amount spent. For example, one patron could buy a $50 pass, while another buys a $100 pass. When the club becomes too crowded, people join the line to wait. However, the first person to enter the nightclub from the line is the one *holding a pass representing the highest payment of anyone in line*. If two or more people in line are tied for having spent the most money, then one of them is chosen to enter the nightclub. Patrons with no pass are treated as having paid $0.

In Figure 4-3, the patron in the middle with a $100 pass is the first one to enter the club, followed by the two $50 patrons (in some order). All other patrons without a pass are considered to be equivalent, and so any one of them could be selected next to enter the club.

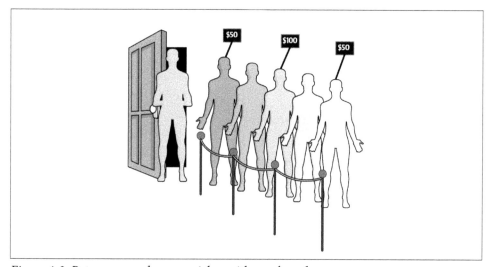

Figure 4-3. Patrons can advance quicker with purchased pass

 A priority queue data type does not specify *what to do when two or more values share the same highest priority*. In fact, based on its implementation, the priority queue might not return values in the order in which they were enqueued. A heap-based priority queue—such as described in this chapter—does not return values with equal priority in the order they were enqueued. The heapq built-in module implements a priority queue using a heap, which I cover in Chapter 8.

This revised behavior defines the *priority queue* abstract data type; however, enqueue() and dequeue() can no longer be efficiently implemented in constant time. On one hand, if you use a linked list data structure, enqueue() would still be O(1), but dequeue() would potentially have to check all values in the priority queue to locate the one with highest priority, requiring O(N) in the *worst case*. On the other hand, if you keep all elements in sorted order by priority, dequeue() requires O(1), but now enqueue() requires O(N) in the *worst case* to find where to insert the new value.

Given our experience to date, here are five possible structures that all use an Entry object to store a (value, priority) entry:

Array
> An *array of unordered entries* that has no structure and hopes for the best. enqueue() is a constant time operation, but dequeue() must search the entire array for the highest priority value to remove and return. Because array has a fixed size, this priority queue could become full.

Built-in
> An *unordered list* manipulated using Python built-in operations that offers similar performance to *Array*.

OrderA
> An *array containing entries sorted by increasing priority*. On enqueue(), use Binary Array Search variation (from Listing 2-4) to locate where the entry should be placed, and manually shift array entries to make room. dequeue() is constant-time because the entries are fully ordered, and the highest priority entry is found at the end of the array. Because the array has a fixed size, this priority queue could become full.

Linked
> A *linked list of entries* whose first entry has highest priority of all entries in the list; each subsequent entry is smaller than or equal to the previous entry. This implementation enqueues new values into their proper location in the linked list to allow dequeue() to be a constant-time operation.

OrderL
> A *Python list containing ascending entries* by increasing priority. On enqueue(), use Binary Array Search variation to dynamically insert the entry into its proper location. dequeue() is constant time because the highest priority entry is always at the end of the list.

To compare these implementations, I devised an experiment that safely performs 3N/2 enqueue() operations and 3N/2 dequeue() operations. For each implementation, I measure the total execution time and divide by 3N to compute the average operation cost. As shown in Table 4-1, a fixed-size array is the slowest, while built-in Python lists cut the time in half. An array of sorted entries halves the time yet again, and a linked list improves by a further 20%. Even so, the clear winner is *OrderL*.

Table 4-1. Average operation performance (time in ns) on problem instances of size N

N	Heap	OrderL	Linked	OrderA	Built-in	Array
256	6.4	2.5	3.9	6.0	8.1	13.8
512	7.3	2.8	6.4	9.5	14.9	26.4
1,024	7.9	3.4	12.0	17.8	28.5	52.9
2,048	8.7	4.1	23.2	33.7	57.4	107.7
4,096	9.6	5.3	46.6	65.1	117.5	220.3
8,192	10.1	7.4	95.7	128.4	235.8	446.6
16,384	10.9	11.7	196.4	255.4	470.4	899.9
32,768	11.5	20.3	–	–	–	–
65,536	12.4	36.8	–	–	–	–

For these approaches, the average cost of an enqueue() or dequeue() operation increases in direct proportion to N. The column, labeled "Heap" in Table 4-1, however, shows the performance results using a Heap data structure; its average cost increases in direct proportion to log(N), as you can see from Figure 4-4, and it significantly outperforms the implementation using ordered Python lists. You know you have logarithmic performance when you only see a constant time increase in runtime performance when the problem size doubles. In Table 4-1, with each doubling, the performance time increases by about 0.8 nanoseconds.

The heap data structure, invented in 1964, provides $O(\log N)$ performance for the operations of a priority queue. For the rest of this chapter, I am not concerned with the values that are enqueued in an entry—they could be strings or numeric values, or even image data; who cares? I am only concerned with the numeric priority for each entry. In each of the remaining figures in this chapter, only the priorities are shown for the entries that were enqueued. Given two entries in a max heap, the one whose priority is larger in value has higher priority.

A heap has a maximum size, M—known in advance—that can store N < M entries. I now explain the structure of a heap, show how it can grow and shrink over time within its maximum size, and show how an ordinary array stores its N entries.

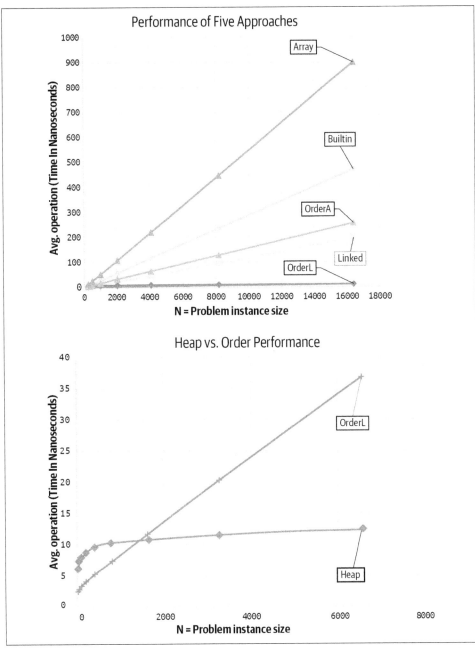

Figure 4-4. O(log N) behavior of Heap outperforms O(N) behavior for other approaches

Max Binary Heaps

It may seem like a strange idea, but what if I only "partially sort" the entries? Figure 4-5 depicts a max heap containing 17 entries; for each entry, *only its priority is shown*. As you can see, level 0 contains a single entry that has the highest priority among all entries in the max heap. When there is an arrow, x → y, you can see that the priority for entry x ≥ the priority for entry y.

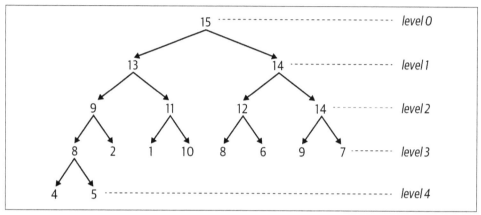

Figure 4-5. A sample max binary heap

These entries are not fully ordered like they would be in a sorted list, so you have to search awhile to find the entry with lowest priority (hint: it's on level 3). But the resulting structure has some nice properties. There are two entries in level 1, one of which must be the second-highest priority (or tied with the highest one, right?). Each level k—*except for the last one*—is *full* and contains 2^k entries. Only the bottommost level is partially filled (i.e., it has 2 entries out of a possible 16), and it is filled from left to right. You can also see that the same priority may exist in the heap—the priorities 8 and 14 appear multiple times.

Each entry has no more than 2 arrows coming out of it, which makes this a *max binary heap*. Take the entry on level 0 whose priority is 15: the first entry on level 1 with priority 13 is its *left child*; the second entry on level 1 with priority 14 is its *right child*. The entry with priority 15 is the *parent* of the two children entries on level 1.

The following summarizes the properties of a max binary heap:

Heap-ordered property
    The priority of an entry is greater than or equal to the priority of its left child and its right child (should either one exist). The priority for each entry (other than the topmost one) is smaller than or equal to the priority of its parent entry.

Heap-shape property

Each level k must be filled with 2^k entries (from left to right) before any entry appears on level $k + 1$.

When a binary heap only contains a single entry, there is just a single level, 0, because $2^0 = 1$. How many levels are needed for a binary heap to store N > 0 entries? Mathematically, I need to define a formula, L(N), that returns the number of necessary levels for N entries. Figure 4-6 contains a visualization to help determine L(N). It contains 16 entries, each labeled using a subscript that starts at e_1 at the top and increases from left to right until there are no more entries on that level, before starting again at the leftmost entry on the next level.

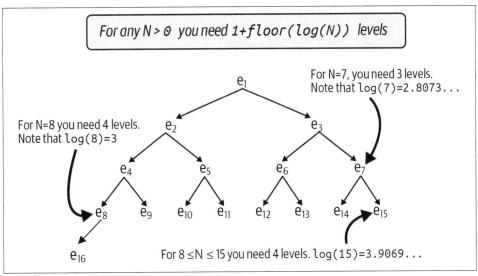

Figure 4-6. Determining levels needed for a binary heap with N entries

If there were only 7 entries in a heap, there would be 3 levels containing entries e_1 through e_7. Four levels would be needed for 8 entries. If you follow the left arrows from the top, you can see that the subscripts follow a specific pattern: e_1, e_2, e_4, e_8, and e_{16}, suggesting that powers of 2 will play a role. It turns out that L(N) = 1 + floor(log(N)).

Each new full level in the binary heap contains more entries than the total number of entries in *all previous levels*. When you increase the height of a binary heap by just one level, the binary heap can contain more than twice as many entries (a total of 2N + 1 entries, where N is the number of existing entries)!

You should recall with logarithms that when you double N, the value of $\log(N)$ increases by 1. This is represented mathematically as follows: $\log(2N) = 1 + \log(N)$. Which of the four options in Figure 4-7 are valid max binary heaps?

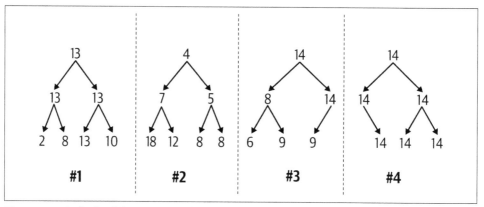

Figure 4-7. Which of these are valid max binary heaps?

First review the heap-shape property for each of these options. Options #1 and #2 are acceptable, since each level is completely full. Option #3 is acceptable because only its last level is partial and it contains three (out of possible four) entries from left to right. Option #4 violates the heap-shape property because its last level contains three entries, but the leftmost possible entry is missing.

Now consider the heap-ordered property for max binary heaps, which ensures that each parent's priority is greater than or equal to the priorities of its children. Option #1 is valid, as you can confirm by checking each possible arrow. Option #3 is invalid because the entry with priority 8 has a right child whose priority of 9 is greater. Option #2 is invalid because the entry with priority 4 on level 0 has smaller priority than both of its children entries.

> Option #2 is actually a valid example of a *min binary heap*, where each parent entry's priority is smaller than or equal to the priority of its children. Min binary heaps will be used in Chapter 7.

I need to make sure that both heap properties hold after `enqueue()` or `dequeue()` (which removes the value with highest priority in the max heap). This is important because then I can demonstrate that both of these operations will perform in $O(\log N)$, a significant improvement to the approaches documented earlier in Table 4-1, which were limited since either `enqueue()` or `dequeue()` had *worst case* behavior of $O(N)$.

Inserting a (value, priority)

After `enqueue(value, priority)` is invoked on a max binary heap, where should the new entry be placed? Here's a strategy that always works:

- Place the new entry into the first available empty location on the last level.
- If that level is full, then extend the heap to add a new level, and place the new entry in the leftmost location in the new level.

In Figure 4-8, a new entry with priority 12 is inserted in the third location on level 4. You can confirm that the heap-shape property is valid (because the entries on the partial level 4 start from the left with no gaps). It might be the case, however, that the heap-ordered property has now been violated.

The good news is that I only need to possibly rearrange entries that lie in the *path* from the newly placed entry all the way back to the topmost entry on level 0. Figure 4-10 shows the end result after restoring the heap-ordered property; as you can see, the entries in the shaded path have been reordered appropriately, in decreasing (or equal) order from the top downward.

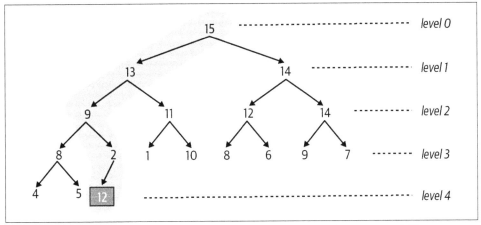

Figure 4-8. The first step in inserting an entry is placing it in the next available position

A *path* to a specific entry in a binary heap is a sequence of entries formed by following the arrows (left or right) from the single entry on level 0 until you reach the specific entry.

To remake the max heap to satisfy the heap-ordered property, the newly added entry "swims up" along this path to its proper location, using pairwise swaps. Based on this example, Figure 4-8 shows that the newly added entry with priority 12 invalidates the heap-ordered priority since its priority is larger than its parent whose priority is 2. *Swap these two entries* to produce the max heap in Figure 4-9 *and continue upward.*

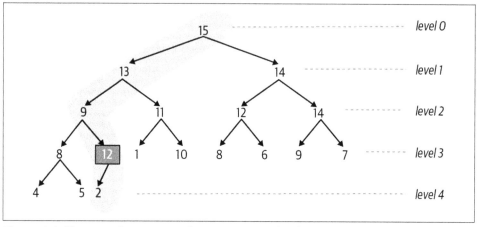

Figure 4-9. The second step swims the entry up one level as needed

You can confirm that from 12 downward, the structure is a valid max binary heap with two entries. However the entry with 12 still invalidates the heap-ordered property since its parent entry has a priority of 9, which is smaller, so swap with its parent, as shown in Figure 4-10.

From highlighted entry 12 downward in Figure 4-10, the structure is a valid max binary heap. When you swapped the 9 and 12 entries, you didn't have to worry about the structure from 8 and below *since all of those values are known to be smaller than or equal to 8*, which means they will all be smaller than or equal to 12. Since 12 is smaller than its parent entry with priority of 13, the heap-ordered property is satisfied.

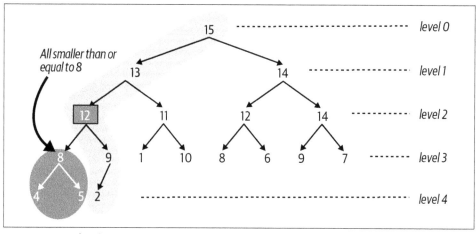

Figure 4-10. Third step swims the entry up one level as needed

Try on your own to enqueue(value, 16) into the heap depicted in Figure 4-10, which initially places the new entry in the fourth location on level 4, as the right child of the entry with priority 9. This new entry will swim up all the way to level 0, resulting in the max binary heap shown in Figure 4-11.

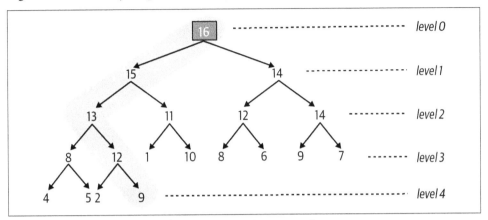

Figure 4-11. Adding entry with priority 16 swims up to the top

The *worst case* is when you enqueue a new entry whose priority is higher than any entry in the max binary heap. The number of entries in the path is 1 + floor(log(N)), which means the *most number of swaps* is one smaller, or floor(log(N)). Now I can state clearly that the time to remake the max binary heap after an enqueue() operation is O(log N). This great result only addresses half of the problem, since I must also ensure that I can efficiently remove the entry with highest priority in the max binary heap.

Removing the Value with Highest Priority

Finding the entry with highest priority in a max binary heap is simple—it will always be the single entry on level 0 at the top. But you just can't remove that entry, since then the heap-shape property would be violated by having a gap at level 0. Fortunately, there is a dequeue() strategy that can remove the topmost entry and efficiently remake the max binary heap, as I show in the next sequence of figures:

1. Remove the rightmost entry on the bottommost level and remember it. The resulting structure will satisfy both the heap-ordered and heap-shape properties.

2. Save the value for the highest priority entry on level 0 so it can be returned.

3. Replace the entry on level 0 with the entry you removed from the bottommost level of the heap. This might break the heap-ordered property.

To achieve this goal, first remove and remember entry 9, as shown in Figure 4-12; the resulting structure remains a heap. Next, record the value associated with the highest priority on level 0 so it can be returned (not shown here).

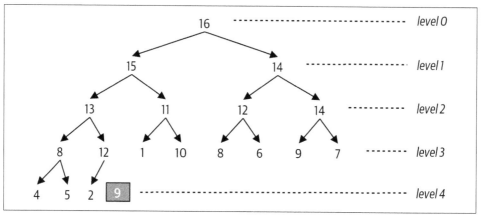

Figure 4-12. First step is to remove bottommost entry

Finally, replace the single entry on level 0 with the entry that was removed, resulting in the broken max heap shown in Figure 4-13. As you can see, the priority of the single entry on level 0 is not greater than its left child (with priority 15) and right child (with priority 14). To remake the max heap, you need to "sink down" this entry to a location further down inside the heap to re-establish the heap-ordered property.

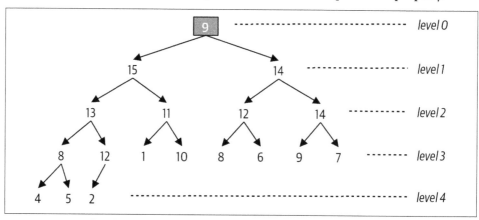

Figure 4-13. Broken heap resulting from swapping last entry with level 0

Starting from the entry that is invalid (i.e., the level 0 entry with priority 9), determine which of its children (i.e., left or right) has the higher priority—if only the left child exists, then use that one. In this running example, the left child with priority 15 has higher priority than the right child with priority 14, and Figure 4-14 shows the result of swapping the topmost entry with the higher selected child entry.

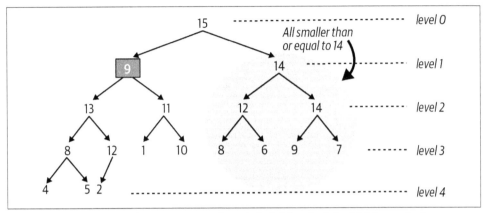

Figure 4-14. Swapping the top entry with its left child, which had higher priority

As you can see, the entire substructure based on the entry with priority 14 on level 1 is a valid max binary heap, and so it doesn't need to change. However, the newly swapped entry (with priority 9) violates the heap-ordered property (it is smaller than the priority of both of its children), so this entry must continue to "sink down" to the left, as shown in Figure 4-15, since the entry with priority 13 is the larger of its two children entries.

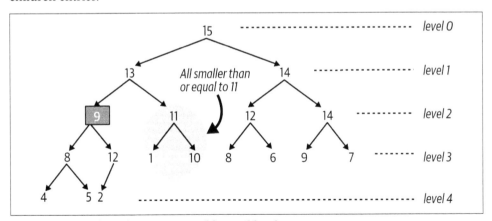

Figure 4-15. Sinking down one an additional level

Almost there! Figure 4-15 shows that the entry with priority 9 has a right child whose priority of 12 is higher, so we swap these entries, which finally restores the heap-ordered property for this heap, as shown in Figure 4-16.

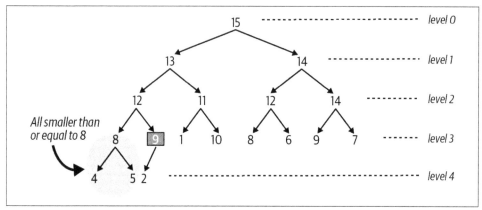

Figure 4-16. Resulting heap after sinking entry to its proper location

There is no simple path of adjusted entries, as we saw when enqueuing a new entry to the priority queue, but it is still possible to determine the maximum number of times the "sink down" step was repeated, namely, one value smaller than the number of levels in the max binary heap, or floor(log(N)).

You can also count the number of times the priorities of two entries were compared with each other. For each "sink down" step, there are at most two comparisons—one comparison to find the larger of the two sibling entries, and then one comparison to determine if the parent is bigger than the larger of these two siblings. In total, this means the number of comparisons is no greater than $2 \times$ floor(log(N)).

It is incredibly important that the max binary heap can both add an entry and remove the entry with highest priority in time that is directly proportional to log(N) in the *worst case*. Now it is time to put this theory into practice by showing how to implement a binary heap using an ordinary array.

Have you noticed that the heap-shape property ensures that you can read all entries in sequence from left to right, from level 0 down through each subsequent level? I can take advantage of this insight by storing a binary heap in an ordinary array.

Representing a Binary Heap in an Array

Figure 4-17 shows how to store a max binary heap of N = 18 entries within a fixed array of size M > N. This max binary heap of five levels can be stored in an ordinary array by mapping each location in the binary heap to a unique index location. Each dashed box contains an integer that corresponds to the index position in the array that stores the entry from the binary heap. Once again, when depicting a binary heap, I only show the priorities of the entries.

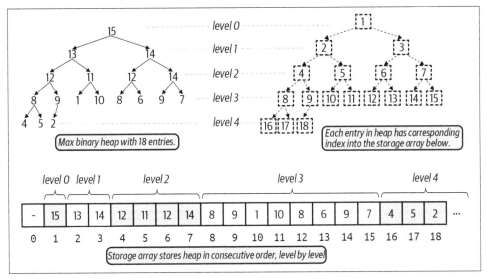

Figure 4-17. Storing a max binary heap in an array

Each entry has a corresponding location in the `storage[]` array. To simplify all computations, location `storage[0]` is unused and never stores an entry. The topmost entry with priority 15 is placed in `storage[1]`. You can see that its left child, with priority 13, is placed in `storage[2]`. If the entry in `storage[k]` has a left child, that entry is `storage[2*k]`; Figure 4-17 confirms this observation (just inspect the dashed boxes). Similarly, if the entry in `storage[k]` has a right child, that entry is in `storage[2*k+1]`.

For k > 1, the parent of the entry in `storage[k]` can be found in `storage[k//2]`, where k//2 is the integer value resulting by truncating the result of k divided by 2. By placing the topmost entry of the heap in `storage[1]`, you just perform integer division by two to compute the parent location of an entry. The parent of the entry in `storage[5]` (with a priority of 11) is found in `storage[2]` because 5//2 = 2.

The entry in `storage[k]` is a valid entry when $0 < k \leq N$, where N represents the number of entries in the max binary heap. This means that the entry at `storage[k]` has no children if $2 \times k > N$; for example, the entry at `storage[10]` (which has priority of 1) has no left child, because $2 \times 10 = 20 > 18$. You also know that the entry at `storage[9]` (which coincidentally has a priority of 9) has no right child, because $2 \times 9 + 1 = 19 > 18$.

Implementation of Swim and Sink

To store a max binary heap, start with an `Entry` that has a `value` with its associated `priority`:

```
class Entry:
  def __init__(self, v, p):
    self.value = v
    self.priority = p
```

Listing 4-2 stores a max binary heap in an array, `storage`. When instantiated, the length of `storage` is one greater than the `size` parameter, to conform to the computations described earlier where the first entry is stored in `storage[1]`.

There are two helper methods that simplify the presentation of the code. You have seen how many times I checked whether one entry has a smaller priority than another entry. The `less(i,j)` function returns `True` when the priority of the entry in `storage[i]` is smaller than the priority of the entry in `storage[j]`. When swimming up or sinking down, I need to swap two entries. The `swap(i,j)` function swaps the locations of the entries in `storage[i]` and `storage[j]`.

Listing 4-2. Heap implementation showing enqueue() *and* swim() *methods*

```
class PQ:
  def less(self, i, j):                              ❶
    return self.storage[i].priority < self.storage[j].priority

  def swap(self, i, j):                              ❷
    self.storage[i],self.storage[j] = self.storage[j],self.storage[i]

  def __init__(self, size):                          ❸
    self.size = size
    self.storage = [None] * (size+1)
    self.N = 0

  def enqueue(self, v, p):                           ❹
    if self.N == self.size:
      raise RuntimeError ('Priority Queue is Full!')

    self.N += 1
    self.storage[self.N] = Entry(v, p)
    self.swim(self.N)

  def swim(self, child):                             ❺
    while child > 1 and self.less(child//2, child):  ❻
      self.swap(child, child//2)                     ❼
      child = child // 2                             ❽
```

❶ `less()` determines if `storage[i]` has lower priority than `storage[j]`.

❷ swap() switches the locations of entries i and j.

❸ storage[1] through storage[size] will store the entries; storage[0] is unused.

❹ To enqueue a (v, p) entry, place it in the next empty location and swim it upward.

❺ swim() remakes the storage array to conform to the heap-ordered property.

❻ The parent of the entry in storage[child] is found in storage[child//2], where child//2 is the integer result of dividing child by 2.

❼ Swap entries at storage[child] and its parent storage[child//2].

❽ Continue upward by setting child to its parent location as needed.

The swim() method is truly brief! The entry identified by child is the newly enqueued entry, while child//2 is its parent entry, should it exist. If the parent entry has lower priority than the child, they are swapped, and the process continues upward.

Figure 4-18 shows the changes to storage initiated by enqueue(value, 12) in Figure 4-8. Each subsequent row corresponds to an earlier identified figure and shows the entries that change in storage. The final row represents a max binary heap that conforms to the heap-ordered and heap-shape properties.

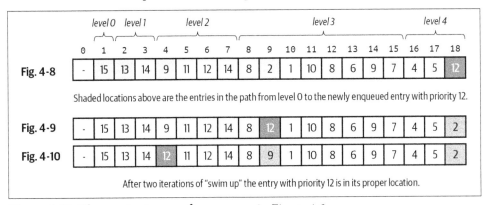

Figure 4-18. Changes to storage after enqueue in Figure 4-8

The path from the topmost entry to the newly inserted entry with priority 12 consists of five entries, as shaded in Figure 4-18. After two times through the while loop in swim(), the entry with priority 12 is swapped with its parent, eventually swimming up to storage[4], where it satisfies the heap-ordered property. The number of swaps will never be more than log(N), where N is the number of entries in the binary heap.

The implementation in Listing 4-3 contains the sink() method to reestablish the structure of the max binary heap after dequeue() is invoked.

Listing 4-3. Heap implementation completed with dequeue() and sink() methods

```
def dequeue(self):
  if self.N == 0:
    raise RuntimeError ('PriorityQueue is empty!')

  max_entry = self.storage[1]                          ❶
  self.storage[1] = self.storage[self.N]               ❷
  self.storage[self.N] = None
  self.N -= 1                                           ❸
  self.sink(1)
  return max_entry.value                                ❹

def sink(self, parent):
  while 2*parent <= self.N:                             ❺
    child = 2*parent
    if child < self.N and self.less(child, child+1):    ❻
      child += 1
    if not self.less(parent, child)                     ❼
      break
    self.swap(child, parent)                            ❽
    parent = child
```

❶ Save entry of highest priority on level 0.

❷ Replace entry in storage[1] with entry from bottommost level of heap and clear from storage.

❸ Reduce number of entries *before* invoking sink on storage[1].

❹ Return the value associated with entry of highest priority.

❺ Continue checking as long as parent has a child.

❻ Select right child if it exists and is larger than left child.

❼ If parent is *not* smaller than child, heap-ordered property is met.

❽ Swap if needed, and continue sinking down, using child as new parent.

Figure 4-19 shows the changes to storage initiated by dequeue() based on the initial max binary heap shown in Figure 4-11. The first row of Figure 4-19 shows the array with 19 entries. In the second row, the final entry in the heap with priority 9 is swapped to become the topmost entry in the max binary heap, which breaks the heap-ordered property; also, the heap now only contains 18 entries, since one was deleted.

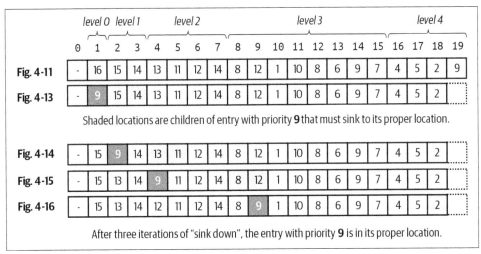

Figure 4-19. Changes to storage after dequeue in Figure 4-11

After three successive passes through the while loop in sink(), the entry with priority 9 has dropped down to a location that ensures the heap-ordered property. In each row, the leftmost highlighted entry is the entry with priority 9, and the shaded entries to the right are its children entries. Whenever the parent entry of 9 is smaller than one of its children, it must sink down to be swapped with the larger of its children. The number of swaps will never be more than log(N).

The sink() method is the hardest to visualize because there is no straight path to follow, as with swim(). In the final representation of storage in Figure 4-19, you can see that the highlighted entry with priority 9 only has one shaded child (with priority 2). When sink() terminates, you know that the entry that was sinking has either reached an index location, p, where it has no children (i.e., because 2 × p is an invalid storage index location greater than N), or it is greater than or equal (i.e., not lesser than) the larger of its children entries.

The order of the statements in dequeue() is *critical*. In particular, you have to reduce N by 1 before calling sink(1), otherwise sink() will mistakenly think the index location in storage corresponding to the recently dequeued entry *is still part of the heap*. You can see in the code that storage[N] is set to None to ensure that entry is not mistakenly thought to be part of the heap.

If you want to convince yourself that the dequeue() logic is correct, consider how it operates with a heap that contains just a single entry. It will retrieve max_entry and set N to 0 before calling sink(), which will do nothing since $2 \times 1 > 0$.

Summary

The binary heap structure offers an efficient implementation of the priority queue abstract data type. Numerous algorithms, such as those discussed in Chapter 7, depend on priority queues.

- You can enqueue() a (value, priority) entry in O(log N) performance.
- You can dequeue() the entry with highest priority in O(log N) performance.
- You can report the number of entries in a heap in O(1) performance.

In this chapter I focused exclusively on max binary heaps. You only need to make one small change to realize a *min binary heap*, where higher priority entries have smaller numeric priority values. This will become relevant in Chapter 7. In Listing 4-2, just rewrite the less() method to use greater-than (>) instead of less-than (<). All other code remains the same.

```
def less(self, i, j):
    return self.storage[i].priority > self.storage[j].priority
```

While a priority queue can grow or shrink over time, the heap-based implementation predetermines an initial size, M, to store the N < M entries. Once the heap is full, no more additional entries can be enqueued to the priority queue. It is possible to automatically grow (and shrink) the storage array, similar to what I showed in Chapter 3. As long as you use geometric resizing, which doubles the size of storage when it is full, then the overall amortized performance for enqueue() remains O(log N).

Challenge Exercises

1. It is possible to use a fixed array, storage, as the data structure to efficiently implement a queue, such that the enqueue() and dequeue() operations have O(1) performance. This approach is known as a *circular queue*, which makes the novel suggestion that the first value in the array isn't always storage[0]. Instead, keep track of first, the index position for the oldest value in the queue, and last, the index position where the next enqueued value will be placed, as shown in Figure 4-20.

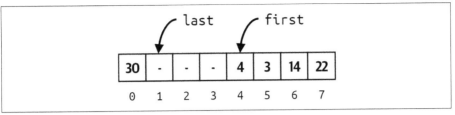

Figure 4-20. Using an array as a circular queue

As you enqueue and dequeue values, you need to carefully manipulate these values. You will find it useful to keep track of N, the number of values already in the queue. Can you complete the implementation in Listing 4-4 and validate that these operations complete in constant time? You should expect to use the modulo % operator in your code.

Listing 4-4. Complete this `Queue` implementation of a circular queue

```
class Queue:
  def __init__(self, size):
    self.size = size
    self.storage = [None] * size
    self.first = 0
    self.last = 0
    self.N = 0

  def is_empty(self):
    return self.N == 0

  def is_full(self):
    return self.N == self.size

  def enqueue(self, item):
    """If not full, enqueue item in O(1) performance."""

  def dequeue(self):
    """If not empty, dequeue head in O(1) performance."""
```

2. Insert N = 2^k – 1 elements in ascending order into an empty max binary heap of size N. When you inspect the underlying array that results (aside from the index location 0, which is unused), can you predict the index locations for the largest k values in the storage array? If you insert N elements in descending order into an empty max heap, can you predict the index locations for *all* N values?

3. Given two max heaps of size M and N, devise an algorithm that returns an array of size M + N containing the combined items from M and N in ascending order

in O(M log M + N log N) performance. Generate a table of runtime performance to provide empirical evidence that your algorithm is working.

4. Use a max binary heap to find the k smallest values from a collection of N elements in O(N log k). Generate a table of runtime performance to provide empirical evidence that your algorithm is working.

5. In a max binary heap, each parent entry has up to two children. Consider an alternative strategy, which I call a *factorial heap*, where the top entry has two children; each of these children has three children (which I'll call grandchildren). Each of these grandchildren has four children, and so on, as shown in Figure 4-21. In each successive level, entries have one additional child. The heap-shape and heap-ordered property remain in effect. Complete the implementation by storing the factorial heap in an array, and perform empirical evaluation to confirm that the results are slower than max binary heap. Classifying the runtime performance is more complicated, but you should be able to determine that it is O(log N/log(log N)).

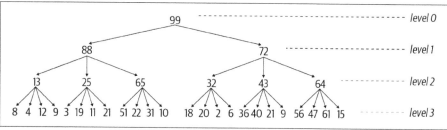

Figure 4-21. A novel factorial heap structure

6. Using the geometric resizing strategy from Chapter 3, extend the PQ implementation in this chapter to automatically resize the storage array by doubling in size when full and shrinking in half when ¼ full.

7. An iterator for an array-based heap data structure should produce the values *in the order they would be dequeued* without modifying the underlying array (since an iterator should have no side effect). However, this cannot be accomplished easily, since dequeuing values would actually modify the structure of the heap. One solution is to create an iterator(pq) generator function that takes in a priority queue, pq, and creates a separate pqit priority queue whose values are index locations in the storage array for pq, and whose priorities are equal to the corresponding priorities for these values. pqit directly accesses the array storage for pq to keep track of the entries to be returned without disturbing the contents of storage.

Complete the following implementation, which starts by inserting into pqit the index position, 1, which refers to the pair in pq with highest priority. Complete the rest of the while loop:

```
def iterator(pq):
  pqit = PQ(len(pq))
  pqit.enqueue(1, pq.storage[1].priority)

  while pqit:
    idx = pqit.dequeue()
    yield (pq.storage[idx].value, pq.storage[idx].priority)


  ...
```

As long as the original pq remains unchanged, this iterator will yield each of the values in priority order.

## CHAPTER 5

Sorting Without a Hat

In this chapter, you will learn:

- How comparison-based sorting algorithms require two fundamental operations:
  - `less(i,j)` determines whether `A[i] < A[j]`.
  - `swap(i,j)` swaps the contents of `A[i]` and `A[j]`.
- How to provide a comparator function when sorting; for example, you can sort integers or string values in descending order. The comparator function can also sort complex data structures with no default ordering; for example, it is not clear how to sort a collection of two-dimensional (x, y) points.
- How to identify inefficient $O(N^2)$ sorting algorithms, such as Insertion Sort and Selection Sort, from the structure of their code.
- *Recursion*, where a function can call itself. This fundamental computer science concept forms the basis of a *divide-and-conquer* strategy for solving problems.
- How Merge Sort and Quicksort can sort an array of N values in $O(N \log N)$ using divide and conquer. How Heap Sort also guarantees $O(N \log N)$.
- How Tim Sort combines Insertion Sort and functionality from Merge Sort to implement Python's default sorting algorithm in guaranteed $O(N \log N)$.

In this chapter, I present algorithms that rearrange the N values in an array so they are in ascending order. Organizing a collection of values in sorted order is an essential first step to improve the efficiency of many programs. Sorting is also necessary for many real-world applications, such as printing staff directories for a company with the names and phone numbers of employees, or displaying flight departure times on an airport display.

With an unordered array, searching for a value, in the *worst case*, is O(N). When the array is sorted, Binary Array Search, in the *worst case*, can locate a target value in O(log N) performance.

Sorting by Swapping

Try sorting the values in the array, A, at the top of Figure 5-1. Use a pencil to copy the values from Figure 5-1 onto a piece of paper (or bring out a pen and just write on these pages!). I challenge you to sort these values in ascending order *by repeatedly swapping the location of two values in the array*. What is the fewest number of swaps that you need? Also, count the number of times you compare two values together. I have sorted these values with five swaps. Is it possible to use fewer?[1]

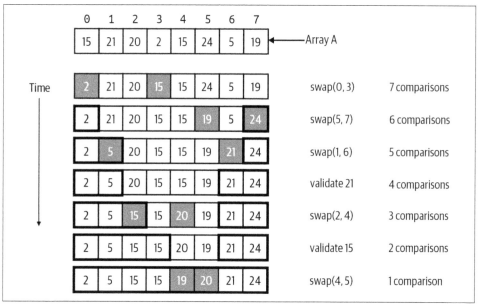

Figure 5-1. Sample array, A, to sort

While it's important to count the number of swaps, you also need to count the *number of comparisons between two values*. To start, you can determine that 2 is the smallest value in A, with just seven comparisons, something I showed in Chapter 1. The smallest value is found at A[3], so it is swapped with A[0]. This moves the smallest value to the front of the array where it belongs. In Figure 5-1, I highlight values when they are swapped. I use bold borders to mark the values that are guaranteed to be in

---

1 No. See the challenge exercises at the end of the chapter.

their final location; these will not be swapped again. All values outside of the bolded borders remain to be sorted.

I scan the remaining values to find the largest value, 24, (using six comparisons) and swap A[5] and A[7] to move the largest value to the end of the array. I then locate the smallest remaining value, 5, (using five comparisons) and swap A[1] and A[6] to move 5 into its proper place. It looks like 21 is in its right spot, which takes four comparisons to validate; no need for a swap here!

With three comparisons, I find that 15 is the smallest remaining value, and I choose to swap the second occurrence of 15, A[4], with A[2]. With two comparisons, you can validate that 15 belongs in index position 3, which leaves just one more comparison to swap A[4] and A[5], moving 19 into its proper spot. In the final step shown in Figure 5-1, the value 20 is in the right location, since it is larger than or equal to all values to its left *and* is smaller than or equal to all values to its right. With five exchanges and 28 comparisons, I have sorted this array.

I didn't follow a specific algorithm to sort this small group of values; sometimes I looked for the smallest value, while other times I looked for the largest. The number of comparisons is reduced after each swap, and there are far more comparisons than swaps. I now define a sorting algorithm that works on any array of N values and evaluate its runtime performance.

Selection Sort

Selection Sort is named because it incrementally sorts an array from left to right, *repeatedly selecting the smallest value remaining* and swapping it into its proper location. To sort N values, find the smallest value and swap it with A[0]. Now only N – 1 values remain to be sorted, since A[0] is in its final spot. Find the location of the smallest remaining value and swap it with A[1], which leaves N – 2 values to sort. Repeat this process until all values are in place.

 What happens when the smallest remaining value is already in its proper place, that is, when i is equal to min_index when the for loop over j completes? The code will attempt to swap A[i] with A[min_index], and nothing in the array will change. You might think to add an if statement to only swap when i and min_index are different, but it will not noticeably improve performance.

In Listing 5-1, there is an outer `for` loop over i that iterates through nearly every index position in the array, from 0 to N − 2. The inner `for` loop over j iterates through the remaining index positions in the array, from i+1 up to N − 1 to find the smallest remaining value. At the end of the `for` loop over i, the value at index position i is swapped with the smallest value found at index position min_index.

Listing 5-1. Selection Sort

```
def selection_sort(A):
  N = len(A)
  for i in range(N-1):            ❶
    min_index = i                 ❷
    for j in range(i+1, N):
      if A[j] < A[min_index]:     ❸
        min_index = j

    A[i],A[min_index] = A[min_index],A[i]   ❹
```

❶ Before each pass through the i `for` loop, you know A[0 .. i-1] is sorted.

❷ min_index is the index location of the smallest value in A[i .. N-1].

❸ If any A[j] < A[min_index], then update min_index to remember index location for this newly discovered smallest value.

❹ Swap A[i] with A[min_index] to ensure that A[0 .. i] is sorted.

At a high level, Selection Sort starts with a problem of size N and reduces it one step at a time, first to a problem of size N − 1, then to a problem of size N − 2, until the whole array is sorted. As shown in Figure 5-2, *it takes N − 1 swaps to sort an array.*

After these swaps have properly placed N − 1 values into their final location, the value at A[N-1] is the largest remaining unsorted value, which means it is already in its final location. Counting the number of comparisons is more complicated. In Figure 5-2, it was 28, which is the sum of the numbers from 1 through 7.

Mathematically, the sum of the numbers from 1 to K is equal to K × (K + 1)/2; Figure 5-3 shows a visualization to provide the intuition behind this formula. Number 28 is called a *triangle number*, from the shape formed by the arrangement of cells.

If you make a second triangle equal in size to the first and rotate it 180 degrees, the two triangles combine to form a K by K + 1 rectangle. The count of the squares in each triangle is half the number of squares in the 7 x 8 rectangle. In this figure, K = 7. When sorting N values, K = N − 1 since that is the number of comparisons in the first step to find the smallest value: the total number of comparisons is (N − 1) × N/2 or ½ × N² − ½ × N.

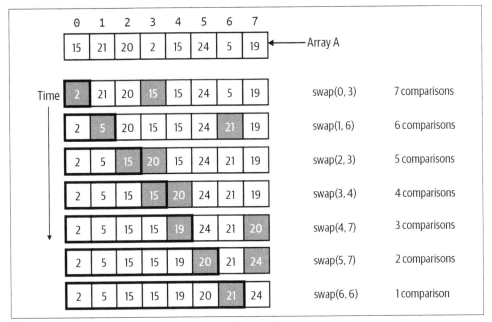

Figure 5-2. Sorting sample array using Selection Sort

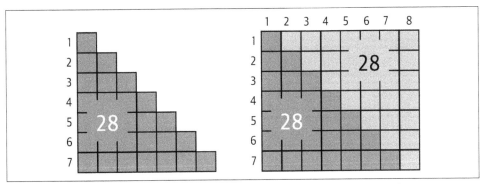

Figure 5-3. Visualizing the formula for triangle numbers: sum of 1 through 7 is 28

Anatomy of a Quadratic Sorting Algorithm

The analysis for Selection Sort shows that the number of comparisons is dominated by the N^2 term, which means its performance will be $O(N^2)$ since that is the dominant operation. To explain why, look at how Selection Sort has $N - 1$ distinct steps when sorting N values. In the first step, it finds the smallest value in $N - 1$ comparisons, and only one value is moved into its proper location. In each of the subsequent $N - 2$ steps, the number of comparisons will (ever so slowly) decrease until there is no work done in the final step. Can something be done to reduce the number of comparisons?

Insertion Sort is a different sorting algorithm that also uses N – 1 distinct steps to sort an array from left to right. It starts by assuming that A[0] is in its proper location (hey, it could be the smallest value in the array, right?). In its first step, it checks if A[1] is smaller than A[0] and swaps these two values as needed to sort in ascending order. In the second step, it tries to *insert* the A[2] value into its proper sorted location when just considering the first three values. There are three possibilities: either A[2] is in its proper spot, or it should be inserted between A[0] and A[1], or it should be inserted before A[0]. However, since you cannot insert a value between two array positions, you must *repeatedly swap values* to make room for the value to be inserted.

At the end of each step, as shown in Figure 5-4, Insertion Sort repeatedly swaps neighboring out-of-order values.

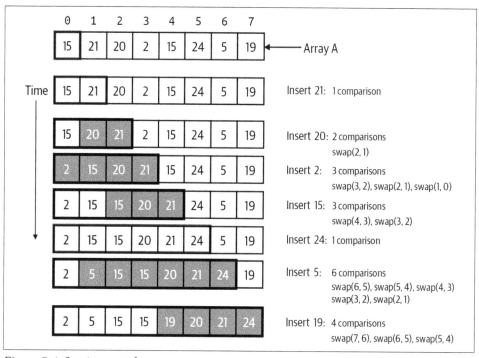

Figure 5-4. Sorting sample array using Insertion Sort

All swapped values are highlighted, and bold borders identify the sorted values in the array. Unlike Selection Sort, values within the bold borders may continue to be swapped, as you can see in the figure. At times (like when the value 5 is inserted) there is a sequence of cascading swaps to move that value into its proper place because the value to insert is smaller than most of the already sorted values. At other times (like when 21 or 24 is inserted), no swaps are needed because the value to insert is larger than all of the already-sorted values. In this example, there are 20 comparisons and 14 swaps. For Insertion Sort, the number of comparisons will always be greater than or equal to the number of swaps. On this problem instance, Insertion Sort uses fewer comparisons than Selection Sort but more swaps. Its implementation, in Listing 5-2, is surprisingly brief.

Listing 5-2. Insertion Sort

```
def insertion_sort(A):
  N = len(A)
  for i in range(1,N):          ❶
    for j in range(i,0,-1):     ❷
      if A[j-1] <= A[j]:        ❸
        break
      A[j],A[j-1] = A[j-1],A[j]  ❹
```

❶ Before each pass through the i for loop, you know A[0 .. i-1] is sorted.

❷ Decrement j from index location i back to 0 but not including 0.

❸ If A[j-1] ≤ A[j], then A[j] has found its proper spot, so stop.

❹ Otherwise, swap these out-of-order values.

Insertion Sort works the hardest when each value to be inserted is smaller than all already-sorted values. The *worst case* for Insertion Sort occurs when the values are in descending order. In each successive step, the number of comparisons (and swaps) increases by one, summing in total to the triangle numbers mentioned earlier.

Analyze Performance of Insertion Sort and Selection Sort

Selection Sort will always have $\frac{1}{2} \times N^2 - \frac{1}{2} \times N$ comparisons and $N - 1$ swaps when sorting N values. Counting the operations for Insertion Sort is more complicated because its performance depends on the order of the values themselves. On average, Insertion Sort should outperform Selection Sort. In the *worst case* for Insertion Sort, the values appear in descending order, and the number of comparisons and swaps is $\frac{1}{2} \times N^2 - \frac{1}{2} \times N$. No matter what you do, both Insertion Sort and Selection Sort require on the order of N^2 comparisons, which leads to the runtime performance visualized in Figure 5-5. Another way to explain this poor behavior is to observe that the problem instance size 524,288 is 512 times as large as 1,024, yet the runtime performance for both Selection Sort and Insertion Sort takes about 275,000 times longer.[2] Sorting 524,288 values takes about two hours for Insertion Sort and nearly four hours for Selection Sort. To solve larger problems, you would need to

---

2 275,000 is about 512 squared.

measure the completion times in days or weeks. This is what a quadratic, or O(N^2), algorithm will do to you, and it is simply unacceptable performance.

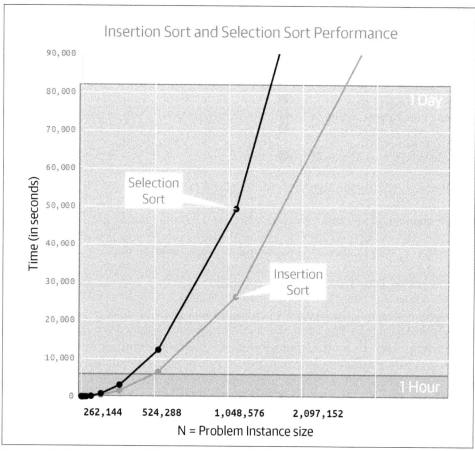

Figure 5-5. Timing results of Insertion Sort and Selection Sort

What if you wanted to sort an array in descending order? Or what if the values have a complex structure and there is no default *less-than* operation defined? Each of the sorting algorithms in this chapter can be extended with a parameter for a comparator function to determine how values are to be ordered, as shown in Listing 5-3. For simplicity, the implementations of the remaining algorithms assume the values are sorted in ascending order.

Listing 5-3. Providing a comparator function to a sorting algorithm

```
def insertion_sort_cmp(A, less=lambda one,two: one <= two):
  N = len(A)
  for i in range(1,N):
    for j in range(i,0,-1):
      if less(A[j-1], A[j]):        ❶
        break
      A[j],A[j-1] = A[j-1],A[j]
```

❶ Determine sorting order using a provided comparator function, less. If less(A[x],A[y]) is True, then A[x] should appear before A[y].

Both Selection Sort and Insertion Sort use N – 1 steps to sort an array of N values, where each step reduces the problem size by just one. A different strategy, known as *divide and conquer*, breaks a problem up into two sub-problems to be solved.

Recursion and Divide and Conquer

The concept of *recursion* has existed in mathematics for centuries—it occurs when a function calls itself.

The *Fibonacci series* starts with two integers, 0 and 1. The next integer in the series is the sum of the two prior numbers. The next few integers in the series are 1, 2, 3, 5, 8, 13, and so on. The recursive formula for the nth integer in the series is F(n) = F(n–1) + F(n–2). As you can see, F(n) is defined by calling itself twice.

The factorial of an integer, N, is the product of all positive integers less than or equal to N. It is written as "N!"; thus 5! = 5 × 4 × 3 × 2 × 1 = 120. Another way to represent this operation is to state that N! = N × (N – 1)! For example, 120 = 5 × 4!, where 4! = 24. A recursive implementation is shown in Listing 5-4.

Listing 5-4. Recursive implementation of factorial

```
def fact(N):
  if N <= 1:            ❶
    return 1
  return N * fact(N-1)  ❷
```

❶ Base case: return 1 for fact(1) or any N ≤ 1.

❷ Recursive case: recursively compute fact(N-1) and multiply its result by N.

It may seem odd to see a function calling itself—how can you be sure that it will not do so forever? Each recursive function has a *base case* that prevents this infinite behavior. fact(1) will return 1 and not call itself.[3] In the *recursive case*, fact(N) calls itself with an argument of N – 1 and multiplies the returned computation by N to produce its final result.

Figure 5-6 visualizes the execution of the statement y = fact(3) as time advances downward. Each box represents an invocation of fact() with the given argument (whether 3, 2, or 1). Invoking fact(3) recursively calls fact(2). When that happens, the original fact(3) function will be "paused" (grayed out in the figure) until the value of fact(2) is known. When fact(2) is invoked, it also must recursively call fact(1), so it is paused (and grayed out in the figure) until the value of fact(1) is known. Finally at this point, the base case stops the recursion, and fact(1) returns 1 as its value, shown inside a dashed circle; this resumes the paused execution of fact(2), which returns 2 × 1 = 2 as its value. Finally, the original fact(3) resumes, returning 3 × 2 = 6, which sets the value of y to 6.

During recursion, any number of fact(N) invocations can be paused until the base case is reached.[4] Then, the recursion "unwinds" one function call at a time until the original invocation is complete.

In reviewing this algorithm, it still solves a problem of size N by reducing it into a smaller problem of size N – 1. What if the problem of size N could be divided into two problems of, more or less, N/2? It might seem like this computation could go on forever, since each of these two sub-problems are further subdivided into four sub-problems of size N/4. Fortunately the base case will ensure that—at some point—the computations will complete.

Consider a familiar problem, trying to find the largest value in an unordered array of N values. In Listing 5-5, find_max(A) invokes a recursive helper function,[5] rmax(0,len(A)-1), to properly set up the initial values for lo = 0 and hi = N – 1, where N is the length of A. The base case in rmax() stops the recursion once lo = hi because this represents looking for the largest value in a range containing just a single value. Once the largest values are determined for the left and right sub-problems, rmax() returns the larger of these two values as the largest value in A[lo .. hi].

---

3 To avoid crashing the Python interpreter because of infinite recursion, this code returns 1 when given any integer less than or equal to 1.

4 In Python, the recursion limit is technically less than 1,000 to prevent crashing the Python interpreter.

5 rmax stands for *recursive max*.

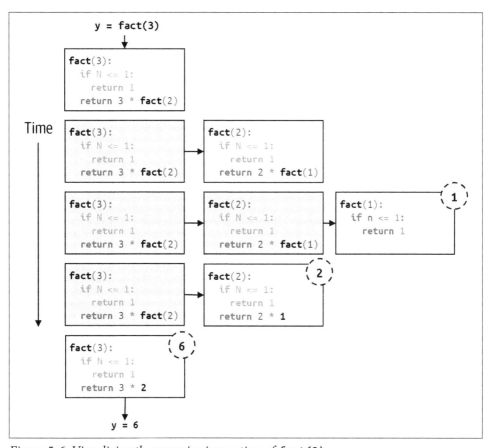

Figure 5-6. Visualizing the recursive invocation of `fact(3)`

Listing 5-5. Recursive algorithm to find largest value in unordered list

```
def find_max(A):

  def rmax(lo, hi):
    if lo == hi: return A[lo]    ❷

    mid = (lo+hi) // 2            ❸
    L = rmax(lo, mid)            ❹
    R = rmax(mid+1, hi)          ❺
    return max(L, R)            ❻

  return rmax(0, len(A)-1)      ❶
```

❶   Invoke the initial recursive call with proper arguments for `lo` and `hi`.

❷ Base case: when `lo == hi`, the range `A[lo .. hi]` contains a single value; return it as the largest value.

❸ Find midpoint index location in the range `A[lo .. hi]`. Use integer division `//` in case range has odd number of values.

❹ L is the largest value in the range `A[lo .. mid]`.

❺ R is the largest value in the range `A[mid+1 .. hi]`.

❻ The largest value in `A[lo .. hi]` is the maximum of L and R.

The function `rmax(lo, hi)` solves this problem recursively by dividing a problem of size N into two problems of half the size. Figure 5-7 visualizes the execution of `rmax(0,3)` on the given array, A, with four values. To solve this problem, it solves two sub-problems: `rmax(0,1)` finds the largest value in the left-hand side of A, and `rmax(2,3)` finds the largest value in the right-hand side of A. Since `rmax()` makes two recursive calls within its function, I introduce a new visualization to describe *where*, in `rmax()`, the execution is paused. I still use a gray background to show that `rmax()` is paused when it makes a recursive call: in addition, the lines highlighted with a black background will execute *once the recursive call returns*.

In Figure 5-7, I only have space to show the three recursive calls that complete to determine that 21 is the largest value in the left-hand side of A. As you can see, the final two lines in the invocation box for `rmax(0,3)` are highlighted in black to remind you that the rest of the computation will resume with the recursive call to `rmax(2,3)`. A similar sequence of three additional recursive calls would complete the right-hand sub-problem, ultimately allowing the original recursive invocation `rmax(0,3)` to return `max(21,20)` as its answer.

Figure 5-8 visualizes the full recursive behavior of `rmax(0,7)`. Similar to my explanation for `fact()`, this figure shows how the invocation of `rmax(0,3)` is paused while it recursively computes the first sub-problem, `rmax(0,1)`. The original problem is repeatedly subdivided until `rmax()` is invoked where its parameters `lo` and `hi` are equal; this will happen eight different times in the figure, since there are N = 8 values. Each of these eight cases represents a base case, which stops the recursion. As you can see in Figure 5-8, the maximum value is 24, and I have highlighted the `rmax()` recursive calls that return this value.

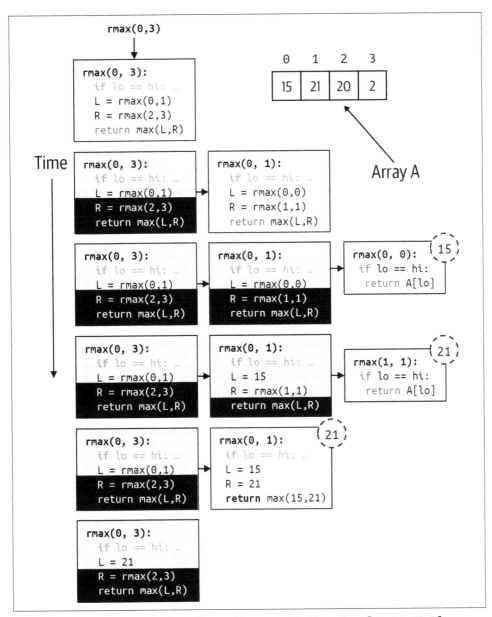

Figure 5-7. Recursive invocation when calling rmax(0,3) on A = [15,21,20,2]

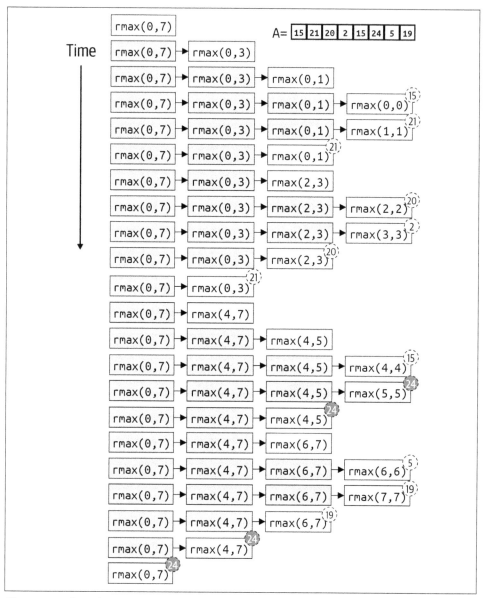

Figure 5-8. Complete recursive invocation of rmax(0, 7)

Merge Sort

Inspired by these examples, we can now ask, "Is there a recursive divide-and-conquer approach to sort an array?" Listing 5-6 contains the gist of an idea: to sort an array, recursively sort its left half, and recursively sort its right half; then somehow *merge the partial results* to ensure the whole array is sorted.

Listing 5-6. Idea for sorting recursively

```
def sort(A):

  def rsort(lo, hi):        ❶
    if hi <= lo: return     ❷

    mid = (lo+hi) // 2
    rsort(lo, mid)          ❸
    rsort(mid+1, hi)
    merge(lo, mid, hi)      ❹

  rsort(0, len(A)-1)
```

❶ Recursive helper method to sort A[lo .. hi].

❷ Base case: a range with one or fewer values is already in sorted order.

❸ Recursive case: sort the left half of A and the right half of A.

❹ Merge both sorted halves of the array in place.

The structure of Listing 5-6 is identical to the find_max(A) function described in Listing 5-5. Completing this implementation leads to Merge Sort, an in-place recursive sorting algorithm that requires extra storage but provides the breakthrough we were looking for, namely an O(N log N) sorting algorithm.

The key to Merge Sort is the merge function that merges *in place* the sorted left half of an array with the sorted right half of an array. The mechanics of merge() might be familiar if you've ever had two sorted stacks of paper that you want to merge into one final sorted stack, as shown in Figure 5-9.

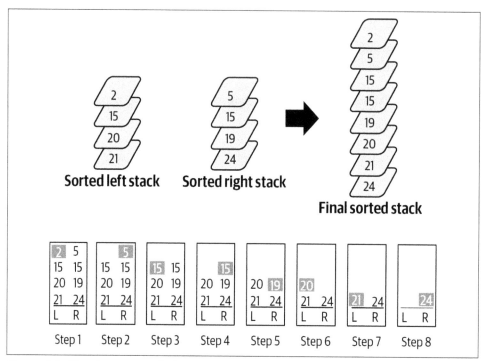

Figure 5-9. Merging two stacks into one

To merge these two stacks into one stack, look at the topmost remaining value in each stack and choose the smallest one. In the first two steps, 2 is removed from the left stack, and then 5 is removed from the right. When faced with two values that are the same, arbitrarily take the value from the left stack, first removing 15 from the left stack, then removing 15 from the right stack. Repeat this process until one of the stacks is exhausted (which happens in the final eighth step). When only one stack remains, just take all those values as a group, since they are already sorted.

The merge process sketched in Figure 5-9 works because of the extra storage into which the values are placed. The most efficient way to implement Merge Sort is to initially allocate extra storage equal to the size of the original array being sorted, as shown in Listing 5-7.

Listing 5-7. Recursive Merge Sort implementation

```
def merge_sort(A):
  aux = [None] * len(A)           ❶

  def rsort(lo, hi):
    if hi <= lo: return           ❷

    mid = (lo+hi) // 2
    rsort(lo, mid)                ❸
    rsort(mid+1, hi)
    merge(lo, mid, hi)

  def merge(lo, mid, hi):
    aux[lo:hi+1] = A[lo:hi+1]     ❹

    left = lo                     ❺
    right = mid+1

    for i in range(lo, hi+1):
      if left > mid:              ❻
        A[i] = aux[right]
        right += 1
      elif right > hi:            ❼
        A[i] = aux[left]
        left += 1
      elif aux[right] < aux[left]: ❽
        A[i] = aux[right]
        right += 1
      else:
        A[i] = aux[left]          ❾
        left += 1

  rsort(0, len(A)-1)              ❿
```

❶ Allocate auxiliary storage equal in size to original array.

❷ Base case: with 1 or fewer values, there is nothing to sort.

❸ Recursive case: sort left and right sub-arrays and then merge.

❹ Copy sorted sub-arrays from A into aux to prepare for merge.

❺ Set left and right to be the starting index positions of the corresponding sub-arrays.

❻ When left sub-array is exhausted, take value from right sub-array.

❼ When right sub-array is exhausted, take value from left sub-array.

❽ When right value is smaller than left value, take value from right sub-array.

❾ When left value is smaller than or equal to right value, take value from left sub-array.

❿ Invoke the initial recursive call.

Figure 5-10 visualizes the dynamic behavior of merge(). The first step of merge(lo,mid,hi) is to copy the elements from A[lo .. hi] into aux[lo .. hi] since this is the sub-problem range being sorted.

The for loop over i will execute 8 times, because that is the total size of the two subproblems being merged. Starting in the third row of Figure 5-10, the variables left, right, and i each keep track of specific locations:

- left is the index position of the next value in the left sub-array to be merged.
- right is the index position of the next value in the right sub-array to be merged.
- i is the index position in A where successively larger values are copied until, by the last step, all values in A[lo .. hi] are in sorted order.

Within the for loop, up to two values in aux (highlighted in Figure 5-10) are compared to find the lower value, which is then copied into A[i]. With each step, i is incremented, while left and right advance only when the value at aux[left] or aux[right] is found to be the next smallest one to be copied into A. The time to complete merge() is directly proportional to the combined size of the sub-problems (or hi - lo + 1).

Merge Sort is a great example of a divide-and-conquer algorithm that guarantees O(N log N) performance. If you have a problem that satisfies the following checklist, then an O(N log N) algorithm exists:

- If you can subdivide a problem of size N into two independent sub-problems of size N/2; it is perfectly fine for one sub-problem to be slightly larger than the other.
- If you have a base case that either does nothing (like with Merge Sort) or performs some operations in constant time.
- If you have a processing step (either before the problem is subdivided or afterward as a post-processing step) that requires time directly proportional to the number of values in the sub-problem. For example, the for loop in merge() repeats a number of times equal to the size of the sub-problem being solved.

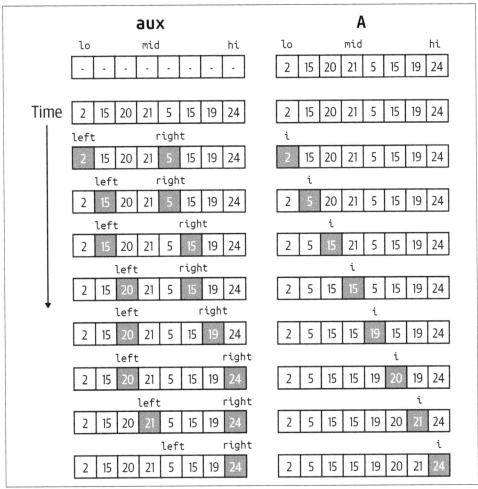

Figure 5-10. Step-by-step merge of two sorted sub-arrays of size 4

Quicksort

Another sorting algorithm that follows divide-and-conquer is Quicksort, one of the most heavily studied and efficient sorting algorithms ever designed.[6] It recursively sorts an array by selecting an element in A to use as a pivot value, p, and then *it inserts p into its proper location in the final sorted array*. To do this, it rearranges the contents of A[lo .. hi] such that there is a left sub-array with values that are ≤ p, and a right

---

6 Invented by Tony Hoare in 1959, Quicksort is well over 50 years old!

sub-array with values that are ≥ p. You can confirm in Figure 5-11 that the partitioned array has this property.

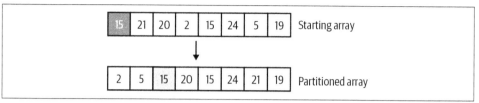

Figure 5-11. Results of `partition(A,0,7,0)` *using* `A[0]` *as pivot*

This amazing feat may at first seem impossible—how do you know where p exists in the final sorted array without actually sorting the entire array? It turns out that partitioning doesn't sort all elements in A but rearranges just a few based on p. In the challenge exercises found in Chapter 1, you can find the implementation of `partition()`. After `partition()` completes in Figure 5-11, the left sub-array to be sorted contains two values, while the right sub-array contains five values. Each of these sub-arrays is recursively sorted using Quicksort, as shown in Listing 5-8.

Listing 5-8. Recursive Quicksort implementation

```
def quick_sort(A):

  def qsort(lo, hi):
    if hi <= lo:                                      ❶
      return

    pivot_idx = lo                                    ❷
    location = partition(A, lo, hi, pivot_idx)        ❸

    qsort(lo, location-1)                             ❹
    qsort(location+1, hi)

  qsort(0, len(A)-1)                                  ❺
```

❶ Base case: with 1 or fewer values, there is nothing to sort.

❷ Choose `A[lo]` as the pivot value, p.

❸ Return `location` in A such that:

- `A[location]` = p

- All values in left sub-array `A[lo .. location-1]` are all ≤ p

- All values in right sub-array `A[location+1 .. hi]` are all ≥ p

❹ Recursive case: sort *in place* left and right sub-arrays, since p is already in its proper sorted location, A[location].

❺ Invoke the initial recursive call.

Quicksort presents an elegant recursive solution whose success depends on the partitioning function. For example, if partition() is invoked on a sub-array A[lo .. hi] containing N values and the smallest value in that sub-array is used as the pivot, then the resulting left sub-array is empty, whereas the right sub-array contains N – 1 values. Reducing a sub-problem by 1 is exactly how Insertion Sort and Selection Sort performed, leading to inefficient $O(N^2)$ sorting. The top of Figure 5-12 summarizes the key steps of Quicksort applied to the array from Figure 5-11. The bottom of Figure 5-12 shows the full recursive execution. On the right side of the figure, you can see A, the array being sorted, and how its values change in response to the recursive execution. For each partition of a range A[lo .. hi], the selected pivot is always A[lo], which is why each box reads partition(lo,hi,lo). As time moves vertically down the figure, you can see how each partition() invocation leads to 1 or 2 recursive calls to qsort(). For example, partition(0,7,0) on A places 15 into its final index location (which is why it is grayed out on the right), leading to two subsequent recursive invocations: qsort(0,1) on the left sub-array and qsort(3,7) on the right sub-array. The invocation of qsort(3,7) does not start until qsort(0,1) has completed its work.

Each time partition is invoked, a different value is placed into its proper index location and grayed out. When qsort(lo,hi) is invoked on a range where lo = hi, that value is in its proper location, and it is also grayed out.

When a partition(lo,hi,lo) produces only a single recursive call to qsort(), it is because the pivot value is placed in either A[lo] or A[hi], thus reducing the problem size by just 1. For example, given the implementation in Listing 5-8, Quicksort will degrade its performance to $O(N^2)$ when called on an array of already-sorted values! To avoid this behavior, Quicksort is often modified to choose the pivot value randomly from within the range A[lo .. hi] by replacing pivot_idx = lo in Listing 5-8 with pivot_idx = random.randint(lo, hi). Decades of research have confirmed that there is always a theoretical possibility that in the *worst case*, Quicksort will have a runtime performance of $O(N^2)$. Despite this weakness, Quicksort is often the sorting algorithm of choice because, unlike Merge Sort, it does not require any extra storage. In reviewing the structure for Quicksort, you can see that it conforms to the checklist for $O(N \log N)$ algorithms.

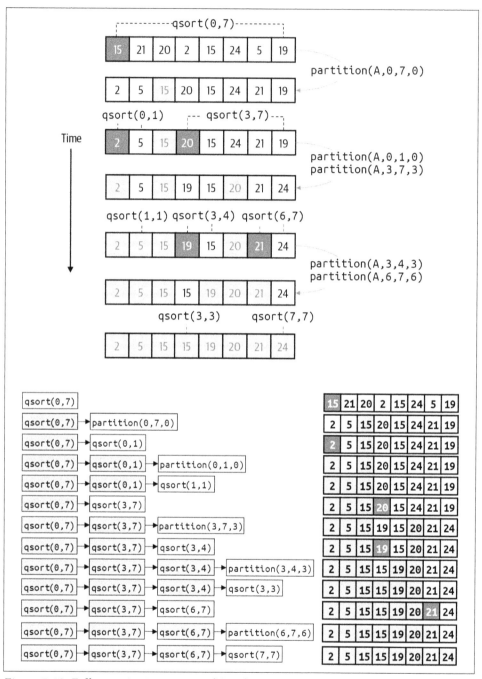

Figure 5-12. Full recursive invocation of Quicksort

Another way to achieve O(N log N) is to have N steps where the runtime perfor-mance of each step is O(log N). Using the heap data structure introduced in the last chapter, I now present Heap Sort, whose runtime performance is O(N log N).

Heap Sort

To see why a max binary heap can help sort an array, consider Figure 5-13 that presents the array storage for the heap from Figure 4-17. The largest value in A is found in A[1]. When this max value is dequeued, the underlying array storage is updated to reflect the modified max binary heap containing one less value. More importantly, the index position A[18] is not only unused, it is *exactly the index posi-tion that should contain the maximum value* if the array were sorted. Simply place the dequeued value there. Perform another dequeue, and this value (the second-largest value in the heap) can be placed in index position A[17], which is now unused.

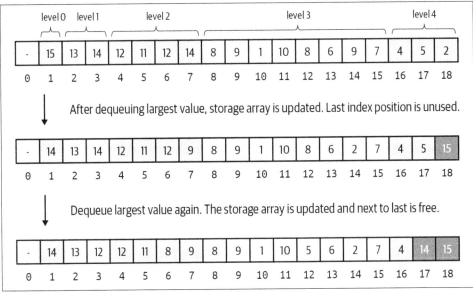

Figure 5-13. Intuition behind how a max binary heap can be used for sorting

To make this promising approach work, I need to address the following issues:

- The heap data structure ignores the value in index position 0 to simplify its com-putations using an array of size N + 1 to store N values.
- The heap is initially empty, and new values are enqueued one at a time. When starting with N values to sort initially, there needs to be an efficient way to "bulk upload" all values.

Let's fix how index positions are calculated. The original heap with 18 elements (as shown in Figure 5-13) was stored in an array with 19 elements. Any reference to A[i] uses 1-based indexing, meaning that A[1] stored the first value in the heap, and A[N-1] stored the last. In Listing 5-9, the less(i,j) and swap(i,j) functions all subtract 1 from i and j whenever accessing A[i] or A[j]. This allows 1-based indexing to work with 0-based array storage. The largest value in the heap is now in A[0]. When swap(1, N) appears in the sort() function, it actually swaps the values in A[0] and A[N-1]. With this small adjustment, the sink() method remains the same. Note that Heap Sort never uses swim().

Listing 5-9. Heap Sort implementation

```
class HeapSort:
  def __init__(self, A):
    self.A = A
    self.N = len(A)

    for k in range(self.N//2, 0, -1):      ❷
      self.sink(k)

  def sort(self):
    while self.N > 1:                      ❸
      self.swap(1, self.N)                 ❹
      self.N -= 1                          ❺
      self.sink(1)                         ❻

  def less(self, i, j):
    return self.A[i-1] < self.A[j-1]       ❶

  def swap(self, i, j):
    self.A[i-1],self.A[j-1] = self.A[j-1],self.A[i-1]
```

❶ To ensure that i // 2 computes the parent index location for i, both less() and swap() subtract 1 from i and j, as if they were using 1-based indexing.

❷ Convert array to be sorted, A, into a max binary heap in bottom-up fashion, starting at N//2, the highest index position that has at least one child.

❸ The while loop continues as long as there are values to sort.

❹ Dequeue maximum value by swapping with last value in heap.

❺ Reduce size of heap by one for upcoming sink() to work.

❻ Sink the newly swapped value into its proper location, which reestablishes the heap-ordered property.

The most important step in Heap Sort is constructing the initial max binary heap from the original array to be sorted. The `for` loop in `HeapSort` completes this task, and the result is shown in Figure 5-14, which required only 23 total comparisons and 5 swaps. This `for` loop constructs a heap from the bottom to the top by starting at index position N//2, the highest index position *that has at least one child*. In reverse order, the `for` loop calls `sink()` on the kth index position to ultimately ensure that all values in the array satisfy the heap-ordered property. These index positions are drawn with a bold border in Figure 5-14.

Through a rather unexpected theoretical analysis, the total number of comparisons required to convert an arbitrary array into a max binary heap is no more than 2N in the *worst case*. The intuition behind this result can be seen in the running total of comparisons in Figure 5-14, which shows a steady, but slow, growth rate. I continue to alternatively shade the index positions within A by the computed level of the max binary heap to show how values are swapped between levels.

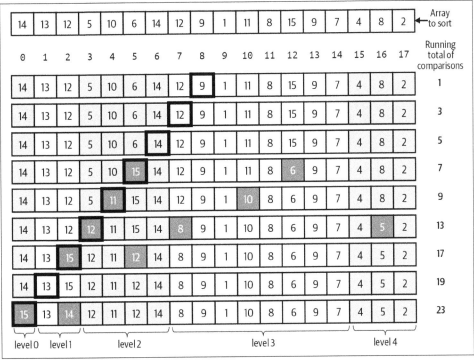

Figure 5-14. Converting array into a max binary heap

The final row in Figure 5-14 represents a max binary heap—in fact, the exact same one depicted in Figure 4-16, now offset by one index position to use all N index positions. The `sort()` function in Listing 5-9 now repeatedly swaps the largest value in the heap with the last value in the heap (using the trick hinted at in Figure 5-13), which has the effect of placing that value in exactly its proper location in the final sorted array. sort() then reduces the size of the heap by one, and sink() properly re-establishes the heap-ordered property with runtime performance of O(log N), as described in Chapter 4.

Performance Comparison of O(N log N) Algorithms

How does the runtime performance of these different sorting algorithms—all classified as O(N log N)—compare with each other? Let's start with some empirical results, as shown in Table 5-1. Reading the numbers down in a column reports the timing results of an algorithm as the problem size doubles; you can see that each timing value is a bit more than twice as large as its previous value. This relative performance is the signature behavior of an O(N log N) algorithm.

Table 5-1. Runtime performance (in seconds) for different sorting algorithms

N	Merge Sort	Quicksort	Heap Sort	Tim Sort	Python Sort
1,024	0.002	0.002	0.006	0.002	0.000
2,048	0.004	0.004	0.014	0.005	0.000
4,096	0.009	0.008	0.032	0.011	0.000
8,192	0.020	0.017	0.073	0.023	0.001
16,384	0.042	0.037	0.160	0.049	0.002
32,768	0.090	0.080	0.344	0.103	0.004
65,536	0.190	0.166	0.751	0.219	0.008
131,072	0.402	0.358	1.624	0.458	0.017
262,144	0.854	0.746	3.486	0.970	0.039
524,288	1.864	1.659	8.144	2.105	0.096
1,048,576	3.920	3.330	16.121	4.564	0.243

Now in each row, the absolute runtime performance of each algorithm is different. In Chapter 2, I discussed how different behaviors within a classification can vary by a multiplicative constant. This table provides evidence of this observation. Once the problem size is large enough, Quicksort is about 15% faster than Merge Sort, while Heap Sort is more than four times slower.

The last two columns in Table 5-1 report on the performance of a new sorting algorithm, Tim Sort, invented by Tim Peters for Python in 2002. This algorithm is quickly becoming the standard sorting algorithm used by major programming languages, such as Java, Python, and Swift. Column "Tim Sort" represents the runtime performance for a simplified Tim Sort implementation, which also exhibits O(N log N) behavior. The final column, labeled "Python Sort," represents the runtime performance using the built-in sort() method in the list data type. Because it is implemented internally, it will naturally be the most efficient—as you can see, it is around 15 times faster than Quicksort. It is worthwhile to investigate Tim Sort because it mixes together two different sorting algorithms to achieve its outstanding performance.

Tim Sort

Tim Sort combines Insertion Sort and the `merge()` helper function from Merge Sort in a novel way to provide a fast sorting algorithm that outperforms other sorting algorithms on real-world data. In particular, Tim Sort dynamically takes advantage of long sequences of partially sorted data to deliver truly outstanding results.

As shown in Listing 5-10, Tim Sort first partially sorts N/`size` sub-arrays of a computed `size`, based on `compute_min_run()`. `size` will typically be an integer between 32 and 64, which means we can treat this number as a constant that is independent of N. This stage ensures there are sequences of partially sorted data, which improves the behavior of `merge()`, the helper function from Merge Sort that merges two sorted sub-arrays into one.

Listing 5-10. Basic Tim Sort implementation

```
def tim_sort(A):
  N = len(A)                                  ❶
  if N < 64:
    insertion_sort(A,0,N-1)
    return

  size = compute_min_run(N)                   ❷
  for lo in range(0, N, size):                ❸
    insertion_sort(A, lo, min(lo+size-1, N-1))

  aux = [None]*N                              ❹
  while size < N:
    for lo in range(0, N, 2*size):
      mid = min(lo + size - 1, N-1)           ❺
      hi  = min(lo + 2*size - 1, N-1)
      merge(A, lo, mid, hi, aux)              ❻

    size = 2 * size                           ❼
```

❶ Small arrays are sorted instead using Insertion Sort.

❷ Compute `size`—a value typically between 32 and 64—to use for the length of the sub-arrays to be sorted.

❸ Use Insertion Sort to sort each sub-array A[lo .. lo+size-1], handling special case when final sub-array is smaller.

❹ merge() uses extra storage equal in size to the original array.

❺ Compute index positions for two sub-arrays to be merged, A[lo .. mid] and A[mid+1 .. hi]. Take special care with partial sub-arrays.

❻ Merge sub-arrays together to sort A[lo .. hi] using aux for auxiliary storage.

❼ Once all sub-arrays of length size are merged with another, prepare for next iteration through while loop to merge sub-arrays twice as large.

The auxiliary storage, aux, is allocated once and used by each invocation of merge(). The actual implementation of Tim Sort has more complicated logic that looks for ascending or strictly descending sub-arrays; it also has a more sophisticated merge function that can merge groups of values "all at once," where the merge() function I've shown operates one value at a time. The simplified implementation whose behavior is shown in Figure 5-15 contains the essential structure. Given the extensive study of sorting algorithms, it is rather amazing that a new sorting algorithm—discovered this century—has proven to be so effective when working with real-world data sets.

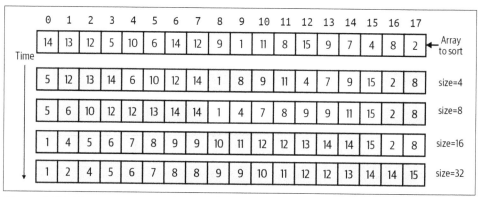

Figure 5-15. Changes to array when applying Tim Sort with initial size of 4

Figure 5-15 demonstrates how Tim Sort works, using a min_run of 4 just to make it easier to visualize. In the first step, four sub-arrays of size 4 are sorted using Insertion Sort; the final two values containing 2 and 8 are contained in a partial sub-array of length 2. These sorted sub-arrays are visualized using alternating bands of shaded and non-shaded regions. There will be N/size sorted sub-arrays (possibly one more, if the length of the original array is not divisible by size). I showed earlier that the run-time performance of sorting size values is directly proportional to size × (size –

1)/2—since this occurs N/size times, the total runtime performance is directly proportional to N × (size − 1)/2. Because size can be considered a constant, this initial phase is classified as O(N).

In the second phase, pairs of neighboring runs are merged together. The total accumulated time for the merge() invocations is proportional to N (as I explained earlier in Merge Sort). After the first pass through the while loop, the size of the sorted sub-arrays has doubled to 8, as you can see by the shaded regions in Figure 5-15. In this example, there are three iterations, as size repeatedly doubles from 4 to 32 (which is greater than N). In general, starting with sorted sub-arrays of size size, the while loop iterates k times until size × 2^k > N; rewrite this computation as 2^k > N/size.

To find k, take the logarithm of both sides, which reveals that k > log(N/size). Because log(a/b) = log(a) − log(b), I can say that k > log(N) − log(size); since size is a constant, I only need to focus on the fact that k is equal to the smallest integer greater than or equal to log(N) minus a small constant value.

To summarize, the first phase of Tim Sort—which applies Insertion Sort—can be classified as O(N), and the second phase—which performs repeated merge() requests— is O(k × N), where k is no greater than log(N), resulting in a total overall performance of O(N log N).

Summary

Sorting is a fundamental problem in computer science and has been extensively studied. An array containing primitive values can be sorted because these values can be compared with each other by default. More complex data types (such as strings or two-dimensional points) can be sorted using custom ordering functions to allow the same sorting algorithms to work.

In this chapter, you learned:

- How some basic sorting algorithms have O(N^2) performance, making them completely unsuitable for sorting large data sets.
- The concept of recursion as a key strategy to solve problems by dividing them into smaller sub-problems.
- That Merge Sort and Heap Sort, in different ways, achieve O(N log N) performance.
- That Quicksort achieves O(N log N) performance without requiring additional storage, as Merge Sort does.
- Tim Sort, the default sorting algorithm used by Python and an increasing number of other programming languages.

Challenge Exercises

1. Write a recursive method count(A,t) that returns the number of times that the value t appears within A. Your implementation must have a recursive structure, similar to find_max(A).

2. You are given an array containing a permutation of the N distinct integers from 0 to N – 1. Determine the fewest number of swaps needed to sort the values in ascending order. Write a function, num_swaps(A), that takes such an array as input and returns an integer value. Note that you do not actually have to sort the array; just determine the number of swaps.

   Extend the problem to work with an array of N distinct values, using the symbol table from Chapter 3, and confirm that five swaps are needed for Figure 5-1.

3. What is the total number of comparisons needed for the recursive find_max(A) to determine the largest value in an unordered array of N values? Is this total less than (or greater than) the total number of comparisons used by largest(A) presented in Chapter 1?

4. In the merge() step in Merge Sort, it can happen that one side (left or right) is exhausted. Currently, the merge() function continues to iterate one step at a time. Replace this logic using Python's ability to copy entire slices of an array, like was done in aux[lo:hi+1] = A[lo:hi+1]. Replace the logic in the first two cases of merge() using slice assignment. Conduct empirical trials to try to measure the performance improvement, if any.

5. Complete a recursive implementation, recursive_two(A), that returns the two largest values in A. Compare its runtime performance against the other approaches from Chapter 1; also compare the number of times less-than is invoked.

6. The Fibonacci series is defined using the recursive formula $F_N = F_{N-1} + F_{N-2}$, with base cases of $F_0 = 0$ and $F_1 = 1$. A related series, *Lucas Numbers*, is defined as $L_N = L_{N-1} + L_{N-2}$, with base cases of $L_0 = 2$ and $L_1 = 1$. Implement fibonacci(n) and lucas(n) using a standard recursive approach and measure the time it takes to compute both F_N and L_N up to N = 40; depending on the speed of your computer, you might have to increase or decrease N to allow the code to terminate. Now implement a new fib_with_lucas(n) method that takes advantage of the following two identities:

   - fib_with_lucas(n): If you set i = n//2 and j = n-i, then $F_{i+j} = (F_i + L_j) \times (F_j + L_i)/2$

   - lucas_with_fib(n): $L_N = F_{N-1} + F_{N+1}$

   Compare timing results of fibonacci() with fib_with_lucas().

## CHAPTER 6

Binary Trees: Infinity in the Palm of Your Hand

In this chapter, you will learn:

- How to create and manipulate binary trees by inserting, removing, and searching for values.

- How to manage a *binary search tree* that enforces the global property that:

  — Values in the left subtree of a node are all smaller than or equal to that node's value.

  — Values in the right subtree of a node are all larger than or equal to that node's value.

- That search, insert, and remove operations can be O(log N) in balanced binary trees but can degrade to become an unacceptable O(N) if you aren't careful.

- How to rebalance a binary search tree after insert and remove operations to guarantee O(log N) performance for search, insert, and remove.

- How to traverse a binary search tree to process each value in ascending order in O(N) performance.

- How to use a binary tree structure to implement a *symbol table* data type with the added benefit that its keys can be retrieved in sorted order.

- How to use a binary tree structure to implement a *priority queue* with the added benefit that you can generate the (key, value) entries in priority order without disrupting the priority queue.

Getting Started

Linked lists and arrays store information in a linear arrangement. In this chapter, I introduce the *binary tree* recursive data structure, one of the most important concepts in the field of computer science. In Chapter 5, you learned about the concept of *recursion*, where a function calls itself. In this chapter, you will learn that the binary tree is a *recursive data structure*—that is, it refers to other binary tree structures. To introduce the concept of a recursive data structure, let's revisit the linked list data structure you have already seen.

A linked list is an example of a recursive data structure because each node has a `next` reference to the first node of a sublist. Linked lists improve upon fixed-length arrays by supporting dynamic growth and reduction of a collection of N values. The `sum_list()` recursive function, shown in Listing 6-1, processes a linked list to return its sum. Compare its implementation against a traditional iterative solution.

Listing 6-1. Recursive and iterative functions to sum the values in a linked list

```
class Node:
  def __init__(self, val, rest=None):
    self.value = val
    self.next = rest

def sum_iterative(n):
  total = 0                              ❶
  while n:
    total += n.value                     ❷
    n = n.next                           ❸
  return total

def sum_list(n):
  if n is None:                          ❹
    return 0
  return n.value + sum_list(n.next)      ❺
```

❶ Initialize `total` to 0 to prepare for computation.

❷ For each node, n, in linked list, add its value to `total`.

❸ Advance to the next node in the linked list.

❹ Base case: the sum of a nonexistent list is 0.

❺ Recursive case: the sum of a linked list, n, is the sum of its value added to the sum of the rest of the list.

The while loop visits each node in the linked list and accumulates a running total of all values stored by nodes in the list. In contrast, sum_list() is a recursive function with a *base case* that terminates the recursion and a *recursive case* that composes results of smaller problem instances together. In the base case, the sum of a nonexistent list is 0. If a list, n, has at least one node, then the recursive case computes the sum of the rest of the list (that is, the list starting with n.next) and adds that result to n.value to determine the total sum.

A linked list of N nodes is recursively decomposed into a first node, n, and the rest, a sublist containing N – 1 nodes. This is a recursive decomposition—since, by definition, the rest is a sublist—but it only subdivides the problem of size N (i.e., find the sum of the values contained in N nodes) into a smaller problem of size N – 1 (i.e., find the sum of the values contained in N – 1 nodes). To envision a recursive data structure that has more productive subdivisions, consider representing basic mathematical expressions using binary operations, such as multiplication. A valid expression is either a value or combines two sub-expressions using a binary operation:

- 3 — any numeric value is valid
- (3 + 2) — add a left value 3 with a right value 2
- (((1 + 5) * 9) – (2 * 6)) — subtract a right expression (2 * 6) from a left expression ((1 + 5) * 9)

Expressions can combine and grow to be as large as desired—the expression in Figure 6-1 has seven mathematical operations and eight numeric values. Linked lists cannot model this non-linear expression. If you've ever tried to visualize genealogies using family trees, then you can see why the diagram is called an *expression tree*.

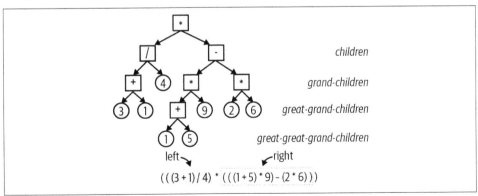

Figure 6-1. Representing mathematical expressions using expression trees

The top multiply node has two child nodes, ultimately leading to four grandchild nodes (one of which is the value 4), six great-grandchild nodes, and two great-great grandchild nodes.

The expression in Figure 6-1 represents the multiplication of two expressions, which demonstrates its recursive structure. To compute the value of this expression, first recursively compute the left expression to produce the result 1. In similar recursive fashion, the right expression evaluates to 42, so the overall result of the original expression is $1 * 42 = 42$.

Figure 6-1 visualizes the recursive substructures of the original expression. At the top is a box representing a multiplication expression, with a left arrow and a right arrow to its left and right sub-expressions. Each circle represents a Value node containing a numeric value, providing the base cases that stop recursion. The Expression data structure in Listing 6-2 models expressions using left and right sub-expressions.

Listing 6-2. Expression data structure to represent mathematical expressions

```
class Value:                                   ❶
  def __init__(self, e):
    self.value = e

  def __str__(self):
    return str(self.value)

  def eval(self):
    return self.value

class Expression:                              ❷
  def __init__(self, func, left, right):
    self.func  = func
    self.left  = left
    self.right = right

  def __str__(self):                           ❸
    return '({} {} {})'.format(self.left, self.func.__doc__, self.right)

  def eval(self):                              ❹
    return self.func(self.left.eval(), self.right.eval())

def add(left, right):                          ❺
  """+"""
  return left + right
```

❶ A Value stores a numeric value. It can return its value and string representation.

❷ An Expression stores a function, func, and left and right sub-expressions.

❸ Provides built-in __str()__ method to recursively produce strings with parentheses around expressions.

❹ Evaluate an Expression by evaluating left and right children and passing those values to func.

❺ Function to perform addition; mult() for multiplication is similar. The doc String __doc__ for the function contains the operator symbol.

Evaluating an Expression is a recursive process that will eventually terminate at Value objects. Using the same technique I introduced in Chapter 5, Figure 6-2 visualizes the recursive evaluation of the expression $m = ((1 + 5) * 9)$:

```
>>> a = Expression(add, Value(1), Value(5))
>>> m = Expression(mult, a, Value(9))
>>> print(m, '=', m.eval())
((1 + 5) * 9) = 54
```

To evaluate m, there could be up to two recursive calls, one on the left and right sub-expressions, respectively. In this case, the left sub-expression, $a = (1 + 5)$ is evaluated recursively, while the right sub-expression, 9, is not. The final computation of 54 is returned as the final result. This example demonstrates the usefulness of the Expression recursive binary tree data structure. It also shows that recursive implementations are brief and elegant.

It is critical that recursive data structures have no structural defects. For example:

```
>>> n = Node(3)
>>> n.next = n              # This is dangerous!
>>> print(sum_list(n))
RecursionError: maximum recursion depth exceeded
```

This linked list is defined by a node, n, whose next node in the linked list is itself! There is nothing wrong with the sum_list() function—the linked list has a structural defect and so the base case for sum_list(n) never occurs. A similar situation could occur in an Expression. These defects are programming mistakes and you can avoid them by carefully testing your code.

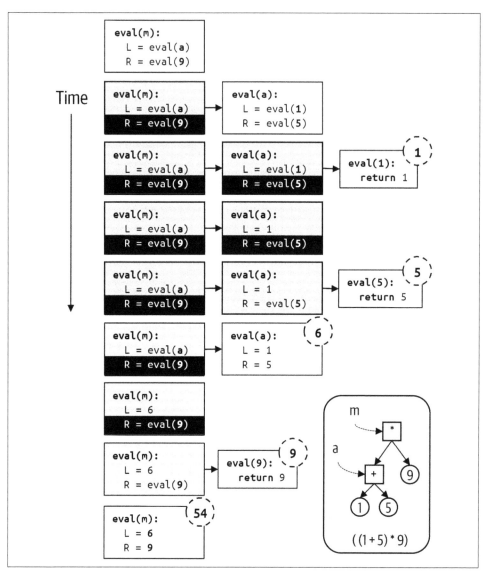

*Figure 6-2. Visualizing recursive evaluation of ((1 + 5) * 9)*

Binary Search Trees

Binary trees are the godfather of all recursive data structures. A *binary search tree* can store collections of values with efficient search, insert, and remove operations.

Storing values in a sorted array is necessary for Binary Array Search to provide $O(\log N)$ performance. There are countless other reasons to produce information in sorted order to make it easier for users to view information. From a practical standpoint, very large fixed arrays are challenging because they require contiguous memory that must be allocated by the underlying operating system. In addition, changing the size of the array is problematic:

- To add a new value to an array, a new and larger array is created, and values from the old array are copied into the new array—while making room for the new value. Finally, memory for the old array is released.
- To remove a value from the array, all values to the right of the removed value must shift one index location to the left. The code must remember that there are "unused" index locations at the end of the array.

Python programs avoid these difficulties because the built-in `list` structure can already grow and shrink without programmer effort, but in the *worst case*, inserting a value into a Python `list` remains $O(N)$. Table 6-1 measures the runtime performance both when prepending 1,000 values (one at a time) to the front of a `list` of size N, and when appending 1,000 values (one at a time) to the end of a `list` of size N.

Table 6-1. Comparing insert and remove performance of lists against binary search tree (time in ms)

N	Prepend	Append	Remove	Tree
1,024	0.07	0.004	0.01	0.77
2,048	0.11	0.004	0.02	0.85
4,096	0.20	0.004	0.04	0.93
8,192	0.38	0.004	0.09	1.00
16,384	0.72	0.004	0.19	1.08
32,768	1.42	0.004	0.43	1.15
65,536	2.80	0.004	1.06	1.23
131,072	5.55	0.004	2.11	1.30
262,144	11.06	0.004	4.22	1.39
524,288	22.16	0.004	8.40	1.46
1,048,576	45.45	0.004	18.81	1.57

As you can see in Table 6-1, the time to append values to the end of a list is constant at 0.004; this can be considered the *best case* for inserting values into a list. The time to prepend 1,000 values to the front of a list essentially doubles when the size of the problem instance, N, doubles. This operation can be classified as O(N). This table demonstrates the hidden cost of using a list to maintain a collection of sorted values. The column labeled "Remove" in Table 6-1 reveals that removing the first value from a list 1,000 times is O(N) too, since its runtime performance doubles as N doubles. Each successive value in this column is almost exactly twice as large as the value above it.

 If the performance of inserting a value 1,000 times is O(N), you know that the performance of inserting a single value is also O(N), using the reasoning I introduced in Chapter 2. Inserting 10,000 values is O(N) using the same reasoning, since these behaviors are all just a multiplicative constant different from each other.

This empirical trial reveals O(N) performance when simply inserting or removing a value; maintaining the array in sorted order will only slow the program down. In contrast, the column labeled "Tree" in Table 6-1 reports the runtime performance when using a *balanced binary search tree* to maintain the collection of values while inserting 1,000 new values. As the problem size doubles, the runtime performance appears to increase by a constant amount, which is characteristic of O(log N) performance. Even better, the binary search tree provides efficient search, insert, and remove operations.

Figure 6-3 contains an example of a *binary search tree* placed side by side with a sorted array containing the same values. This is the same array from Figure 2-5, where I presented Binary Array Search. The top node in a binary tree is designated as the *root* of the tree, analogous to how the *first* node in a linked list is specially designated. In total there are seven nodes in this tree.

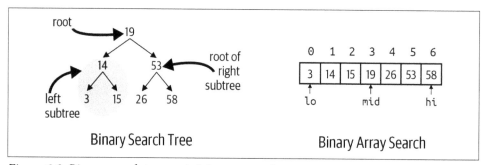

Figure 6-3. Binary search tree containing seven values

Each node in the binary search tree has the structure defined in Listing 6-3. `left` refers to a node that is the root of its own subtree; the same is true of `right`. A binary search tree adds two global constraints for each node, n, in the tree:

- If a node, n, has a `left` subtree, all values in the subtree are $\leq$ n.value.
- If a node, n, has a `right` subtree, all values in the subtree are $\geq$ n.value.

You can confirm these properties hold in Figure 6-3. A *leaf* node is a node without a `left` or `right` subtree; there are four *leaf* nodes in this tree, containing the values 3, 15, 26, and 58. Many people have commented that computer science trees are upside down, because the leaves are at the bottom, while the root is at the top.

Listing 6-3. Structure of a binary search tree

```
class BinaryNode:
  def __init__(self, val):
    self.value = val        ❶
    self.left  = None       ❷
    self.right = None       ❸
```

❶ Each node stores a `value`.

❷ Each node's `left` subtree, if it exists, contains values $\leq$ `value`.

❸ Each node's `right` subtree, if it exists, contains values $\geq$ `value`.

Referring back to Figure 6-3, you can see how the left subtree of the root node is itself a tree whose root node, 14, has a left leaf node (representing 3) and a right leaf node (representing 15). These are exactly the same values in the array depicted on the right that are smaller than or equal to 19, the middle index in the array. A binary tree grows top to bottom as values are inserted, one at a time, as shown in Table 6-2.

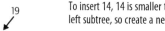

19	To insert 19, create a new subtree with root of 19.
	To insert 14, 14 is smaller than or equal to 19, so insert 14 into the left subtree of 19, but there is no left subtree, so create a new subtree with root of 14.
	To insert 15, 15 is smaller than or equal to 19, so insert 15 into the left subtree of 19 rooted at 14. Now 15 is larger than 14, so insert 15 into the right subtree of 14, but there is no right subtree, so create a new subtree with root of 15.
	To insert 53, 53 is larger than 19, so insert 53 into the right subtree of 19, but there is no right subtree, so create a new subtree with root of 53.
	To insert 58, 58 is larger than 19, so insert 58 into the right subtree of 19 rooted at 53. Now 58 is larger than 53, so insert 58 into the right subtree of 53, but there is no right subtree, so create a new subtree with root of 58.
	To insert 3, 3 is smaller than or equal to 19, so insert 3 into the left subtree of 19 rooted at 14. Now 3 is smaller than or equal to 14, so insert 3 into the left subtree of 14, but there is no left subtree, so create a new subtree with root of 3.
	To insert 26, 26 is larger than 19, so insert 26 into the right subtree of 19 rooted at 53. Now 26 is smaller than or equal to 53, so insert 26 into the left subtree of 53, but there is no left subtree, so create a new subtree with root of 26.

It is convenient to have a `BinaryTree` class to maintain the reference to the `root` node for a binary tree; over time in this chapter, additional functions will be added to this class. Listing 6-4 contains the code needed to insert values into a binary search tree.

Listing 6-4. `BinaryTree` class to improve usability of binary search tree

```
class BinaryTree:
  def __init__(self):
    self.root = None                                ❶

  def insert(self, val):                            ❷
    self.root = self._insert(self.root, val)

  def _insert(self, node, val):
    if node is None:
      return BinaryNode(val)                        ❸

    if val <= node.value:                           ❹
      node.left = self._insert(node.left, val)
    else:                                           ❺
      node.right = self._insert(node.right, val)
    return node                                     ❻
```

❶ `self.root` is the root node of the `BinaryTree` (or `None` if empty).

❷ Use `_insert()` helper function to insert `val` into tree rooted at `self.root`.

❸ Base case: to add `val` to an empty subtree, return a new `BinaryNode`.

❹ If `val` is smaller than or equal to `node`'s value, set `node.left` to be the subtree that results when inserting `val` into subtree `node.left`.

❺ If `val` is larger than `node` value, set `node.right` to be the subtree that results when inserting `val` into subtree `node.right`.

❻ This method *must return* `node` to uphold its contract that it returns the root of the subtree into which `val` was inserted.

The `insert(val)` function in `BinaryTree` invokes the recursive `_insert(node, val)` helper function to set `self.root` to be the subtree that results when inserting `val` into the subtree rooted at `self.root`.[1]

The casual and elegant one-line implementation of `insert()` is a feature of programs using recursive data structures. The `_insert()` function both inserts `val` and returns the root of the resulting subtree.

---

1 All recursive helper functions start with an underscore (_) to declare that these functions are not intended to be part of the public interface for `BinaryTree`.

While all of the values being inserted are unique in this example, in general, binary search trees can contain duplicate values, which is why the _insert() function checks if val ≤ node.value.

In _insert(node, val), the base case occurs when node is None, which occurs whenever val is requested to be inserted into a nonexistent subtree; it just returns a new subtree rooted by the newly created BinaryNode. For the recursive case, val is inserted into either the left subtree, node.left, or the right subtree, node.right. At the end of the recursive case, it is critical that _insert() returns node to fulfill its obligation of returning the root of the subtree that results from adding val to the subtree rooted at node.

_insert(node, val) maintains the binary search tree property such that all values in the left subtree for node are smaller than or equal to node.value, and values in the right subtree are larger than or equal to node.value.

 A node, n, in a binary tree can have a left and a right child. This makes n the *parent node* for left and right. The *descendants* of n are the nodes in its left and right subtrees. Each node, other than the root, has at least one ancestor from which it descends.

Try inserting 29 into the binary search tree from Figure 6-3. 29 is larger than the root, so it must be inserted into the right subtree rooted by 53. 29 is smaller than 53, so insert into its left subtree rooted at 26. Finally, 29 is larger than 26, so it forms the new right subtree for 26, as shown in Figure 6-4.

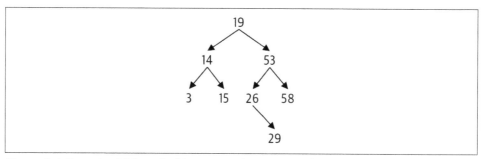

Figure 6-4. Inserting 29 into the binary search tree example

The order in which values are inserted determines the structure of the final binary tree, as you can see in Figure 6-5. For the binary search tree on the left, you know that 5 was the first value inserted, since it is the root of the tree. In addition, every node must have been inserted after its ancestor.

The binary search tree on the right was formed by inserting the seven values in increasing order, which reveals the *worst case* for inserting values into a binary search tree; if you rotate the image counterclockwise about 45 degrees, it looks like a linked list, in which case it loses its efficiency. Toward the end of this chapter, I present a strategy to maintain more balanced tree structures in the face of insertions and removals.

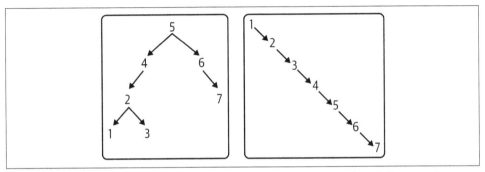

Figure 6-5. Different binary search trees when values are inserted in different order

Searching for Values in a Binary Search Tree

The _insert() method recursively finds the appropriate location to insert a new leaf node containing the value to be added. This same recursive approach could simply check whether a value is contained in a binary search tree; in practice, however, the code in Listing 6-5 offers a simpler, non-recursive solution using a while loop.

Listing 6-5. Determining whether a BinaryTree contains a value

```
class BinaryTree:
  def __contains__(self, target):
    node = self.root                    ❶
    while node:
      if target == node.value:          ❷
        return True

      if target < node.value:           ❸
        node = node.left
      else:
        node = node.right               ❹

    return False                        ❺
```

❶ Start the search at the root.

❷ If target value is same as node's value, return True for success.

❸ If `target` is smaller than `node`'s `value`, set `node` to its `left` subtree to continue search in that subtree.

❹ If `target` had been larger than `node`'s value, continue search in `right` subtree.

❺ If the search runs out of nodes to inspect, the value does not exist in the tree, so return `False`.

This `__contains__()` function is added to the `BinaryTree` class.[2] Its structure is similar to searching for a value within a linked list; the difference is that the next node to be searched could either be `left` or `right` based on the relative value of `target`.

Removing Values from a Binary Search Tree

Removing a value from a linked list was relatively straightforward—as discussed in Chapter 3—but removing a value from a binary search tree is more challenging. To start with, if the value to be removed is contained in the root node, how do you "stitch together" its orphaned left and right subtrees? Also, there should be a least-effort consistent strategy that works every time. Let's try to find an intuitive solution to remove the value contained in the root node of a binary search tree. Figure 6-6 offers two possible binary search trees after removing the root value of 19.

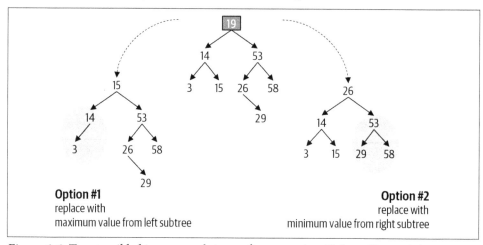

Figure 6-6. Two possible binary search trees after removing 19 from Figure 6-4

---

2 Implementing this function means a program can use the Python `in` operator to determine if a value is contained in a `BinaryTree` object.

You can confirm that both of these options remain binary search trees: the values in each left subtree remain smaller than or equal to its root, and the values in each right subtree remain larger than or equal to its root. The effort involved appears to be minimal for both options:

- Option #1: Find and remove the maximum value in the left subtree, and use that value for the root.
- Option #2: Find and remove the minimum value in the right subtree, and use that value for the root.

Each of these options is acceptable, and I choose to implement the second one. The resulting binary tree is a valid binary search tree because the new root value of 26 is the smallest value in the original right subtree—which means by definition it is smaller than or equal to all values in the revised subtree shaded in Figure 6-6. In addition, it is larger than or equal to all values in the original left subtree because it is larger than or equal to the original root value of 19, which was already larger than or equal to the values in the original left subtree.

Let's start by solving the sub-problem that removes the minimum value in a given subtree. If you think about it, the minimum value in a subtree *cannot have a left child* —since otherwise a smaller value would exist. Given the binary search tree in Figure 6-7, the minimum value in the right subtree rooted at 53 is 26, and as you can see, it has no left child. Removing this value only requires "lifting up" its right subtree, rooted at 29, to become the new left subtree to 53. Setting the left child of 53 to be the tree rooted at 29 will always work, since 26 *has no left subtree* and no values will be lost.

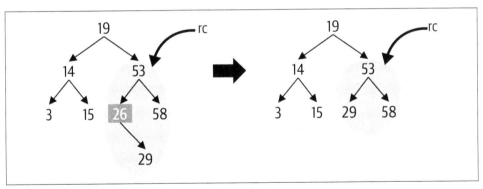

Figure 6-7. Removing minimum value in a subtree

Listing 6-6 contains the helper function in BinaryTree, _remove_min(node), that removes the minimum value in the subtree rooted at node; this function is never called when node is None. When invoking _remove_min() on rc—the right child of the tree in Figure 6-7—the recursive case is invoked, namely to remove the minimum value in the left subtree rooted at 26. This leads to the base case, since 26 has no left subtree, and in its place the function "lifts up" and returns its right subtree rooted at 29 to become the new left subtree to 53.

Listing 6-6. Removing minimum value

```
def _remove_min(self, node):
  if node.left is None:              ❶
    return node.right

  node.left = self._remove_min(node.left)  ❷
  return node                        ❸
```

❶ Base case: if node has no left subtree, then *it is the smallest value* in the subtree rooted at node; to remove it, just "lift up" and return its right subtree (which could be None).

❷ Recursive case: remove the minimum value from left subtree, and the returned subtree becomes new left subtree for node.

❸ _remove_min() completes the recursive case by returning the node whose left subtree may have been updated.

Once again, this code is brief and elegant. As with the other recursive functions discussed earlier, _remove_min() returns the root node of the modified subtree. With this helper function, I can now complete the implementation of remove() that removes a value from a binary search tree. To visualize what the code must do, Table 6-3 shows an example of the changes to a binary tree when removing the value, 19, contained in its root node.

Table 6-3. Demonstrating how root node is removed from binary search tree

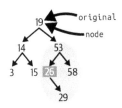

Since the root node contains the value to be removed, set `original` to be the same as node.

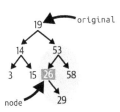

Once the `while` loop completes, node is changed to refer to the *smallest* value in the `right` subtree of `original`—in this case, the node whose value contains 26. This is going to be the new root for the entire subtree. It is important to see that (a) node has no `left` subtree, (b) its value is the smallest value in the subtree rooted by 53, and (c) it is larger than or equal to all values in the subtree rooted by 14.

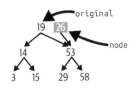

After removing the minimum value from the subtree rooted at `original.right` (containing value 53), `node.right` is set to this updated subtree (which consists of the three nodes with values 29, 53, and 58). For a brief moment, `original.right` and `node.right` both point to the subtree rooted at 53.

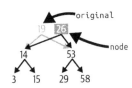

To complete the update, `node.left` is set to refer to `original.left`. When `_remove()` is done, it returns node, which will then "take the place" of `original`, whether as the root node for the entire binary search tree or as a child node to another node.

The implementation of `remove()` is shown in Listing 6-7.

Listing 6-7. Removing a value from a `BinaryTree`

```
def remove(self, val):
  self.root = self._remove(self.root, val)          ❶

def _remove(self, node, val):
  if node is None: return None                       ❷

  if val < node.value:
    node.left = self._remove(node.left, val)         ❸
  elif val > node.value:
    node.right = self._remove(node.right, val)       ❹
  else:                                              ❺
    if node.left is None:  return node.right
    if node.right is None: return node.left          ❻

    original = node                                  ❼
    node = node.right
    while node.left:                                 ❽
      node = node.left

    node.right = self._remove_min(original.right)    ❾
    node.left = original.left                         ❿

  return node
```

❶   Use `_remove()` helper function to remove `val` from tree rooted at `self.root`.

❷   Base case: attempting to remove `val` from nonexistent tree returns `None`.

❸   Recursive case #1: if value to be removed is smaller than `node.value`, set `node.left` to be the subtree that results from removing `val` from `node.left`.

❹   Recursive case #2: if value to be removed is larger than `node.value`, set `node.right` to be the subtree that results from removing `val` from `node.right`.

❺   Recursive case #3: it may be that `node` is root of subtree and contains value to be removed, so there's work to be done.

❻   Handle easy cases first. If `node` is a leaf, then `None` is returned. If it has just one child, then return that child node.

❼   Remember original reference to `node`, since we don't want to lose track of `node`'s original `left` and `right` subtrees, both of which must exist.

❽   Start with `node = node.right` to find the smallest value in the subtree rooted at `node.right`: as long as `node` has a left subtree, then it does not contain the smallest value, so iteratively locate the node with no left subtree—this is the smallest value in the right subtree of original.

❾ node will become the new root to the left and right children of original. Here I set node.right to the subtree that results from removing the minimum value from original.right. You might notice that this recursive method essentially repeats the process of the while loop, but this code is much easier to understand than trying to do everything in just one pass.

❿ Stitch the subtree rooted at node back together.

The final capability supported by a binary search tree is returning the values in ascending order. In computer science this is known as a *traversal*.

Traversing a Binary Tree

To process each element in a linked list, start at the first node and use a while loop to follow next references until all nodes are visited. This linear approach can't work with a binary search tree because there are left and right references to follow. Given the recursive nature of a binary tree data structure, there needs to be a recursive solution. Listing 6-8 contains an elegant recursive solution that uses Python generators.

Listing 6-8. Generator that iterates over values in binary search tree in ascending order

```
class BinaryTree:

  def __iter__(self):
    for v in self._inorder(self.root):      ❶
      yield v

  def _inorder(self, node):
    if node is None:                        ❷
      return

    for v in self._inorder(node.left):      ❸
      yield v

    yield node.value                        ❹

    for v in self._inorder(node.right):     ❺
      yield v
```

❶ Yield all values that result from the *in order* traversal of binary search tree rooted at self.root

❷ Base case: nothing to generate for a nonexistent subtree.

❸ To generate all values in order, first generate all values in order from the subtree rooted at node.left.

❹ Now it is node's turn to yield its value.

❺ Finally, generate all values in order from the subtree rooted at node.right.

The __iter()__ function repeatedly yields the values provided by the recursive helper function, _inorder(), using a common idiom provided by Python. For the base case of the recursion, when asked to yield the values for a nonexistent binary search tree rooted at node, _inorder() returns and does nothing. For the recursive case, this function relies on the binary search tree property that all values in the subtree rooted at node.left are smaller than or equal to node.value and that all values in the subtree rooted at node.right are larger than or equal to node.value. It recursively yields all values in node.left before yielding its own value, and subsequently yielding the values in node.right. The process is visualized in Figure 6-8 with a binary search tree, T, containing five values.

 You can also choose to traverse a binary tree in two other traversal strategies. Use a *preorder* traversal to copy a binary tree. A *postorder* traversal visits all children before the parent, so use it to evaluate the value of an expression tree, such as shown in Figure 6-1.

I have now shown how to search, insert, and remove values from a binary search tree; in addition, you can retrieve these values *in ascending order*. It's time to collectively analyze the performance of these fundamental operations.

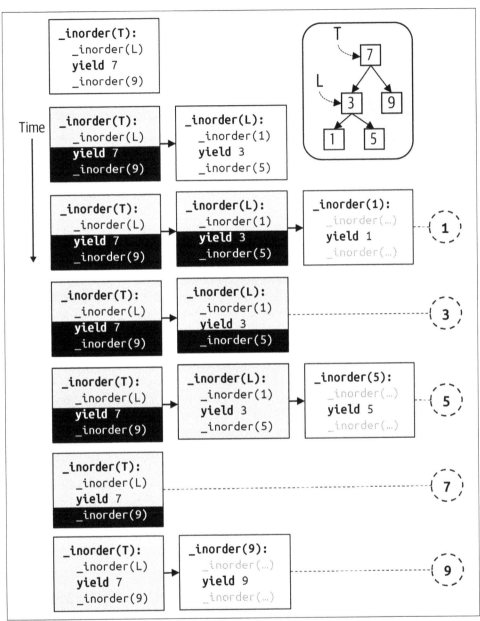

Figure 6-8. Iterating over the values in a binary search tree in ascending order

Analyzing Performance of Binary Search Trees

The determining factor for search, insert, and remove operations is the *height* of the tree, which is defined as the height of its root node. The height of a node is the number of `left` or `right` references you need to get from that node to its most distant descendant leaf node. This means the height of a leaf node is 0.

> The height of a nonexistent binary node cannot be 0, since leaf nodes have a height of 0. The height of None—that is, a nonexistent binary node—is defined as –1 to make computations consistent.

In the *worst case*, the number of nodes visited during a search is based on the height of the root of the binary search tree. Given N nodes in a binary search tree, what is its height? It depends entirely on the order in which the values had been inserted. A *complete binary tree* represents the *best case* since it efficiently stores N = 2^k – 1 nodes in a tree whose height is k – 1. The binary tree in Figure 6-9, for example, has N = 63 nodes, and the height of the root node is 5. Searching for a `target` value will involve no more than 6 comparisons (since with 5 `left` or `right` references you will visit 6 nodes). Since 2^6 – 1 = 63, this means that the time to search for a value is proportional to log (N + 1). But in the *worst case*, all values were inserted in ascending (or descending) order, and the binary search tree is just a long linear chain, as shown in Figure 6-5. In general, the runtime performance of search is O(h), where h is the height of the binary search tree.

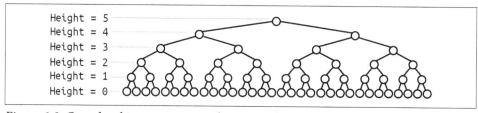

Figure 6-9. Complete binary tree stores the most values with the least height

Inserting a value has the same time complexity as searching for a value—the only difference is that a new leaf node is inserted once the search ends with a nonexistent `left` or `right` subtree; so inserting a value is O(h) as well.

Removing a value from a binary search tree requires three steps:

1. Locate the node containing the value to be removed

2. Locate the minimum value in the right subtree of the node containing the value to be removed

3. Remove that value from the right subtree

Each of these substeps in the *worst case* can be directly proportional to the height.[3] At worst, then, the time to remove a value is proportional to $3 \times h$, where h is the height of the binary search tree. Based on the results from Chapter 2, since 3 is just a multiplicative constant, this means that the time to remove a value remains $O(h)$.

The structure of the binary search tree is based entirely on the order in which the values are inserted and removed. Since the binary search tree cannot control how it is used, it needs a mechanism to detect when its structure is performing poorly. In Chapter 3, I explained how hashtables resized themselves—rehashing all its entries— once a threshold size was hit. It was acceptable for hashtables to do this, because this costly $O(N)$ operation would become ever more infrequently requested using the geometric resizing strategy. As you may recall, doing so enabled the average $O(1)$ runtime performance for get().

The geometric resizing strategy will not work here because there is no simple threshold computation based on N that can determine when to resize, and you can't ensure that resize events become ever more infrequent: all it takes is a small sequence of awkward insertions to unbalance a tree, as visualized in Figure 6-10. Each node is color-coded by its height.

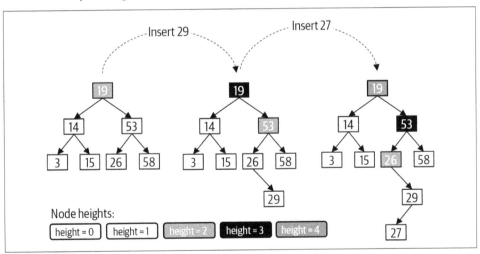

Figure 6-10. Unbalanced tree after two insertions

The complete binary tree on the left in Figure 6-10 has a perfectly balanced structure; each leaf node has a height of 0, and the root node has a height of 2. When 29 is

---

3 See a challenge exercise at the end of the chapter on this point.

inserted—as shown in the middle tree—a new leaf node is created in the proper location. Note that *all ancestor nodes of 29 increase their height by 1*, and they are shaded accordingly. After 27 is inserted—as shown in the right tree—the tree has lost its balance: its left subtree at 14 has a height of 1, but its right subtree at 53 has a height of 3. Other nodes—such as 26 and 53—are similarly out of balance. In the next section, I explain a strategy for detecting and rebalancing binary search trees.

Self-Balancing Binary Trees

The first known self-balancing binary tree data structure, the AVL tree, was invented in 1962.[4] The premise is that as values are inserted into, or removed from, a binary search tree, weaknesses in the structure of the resulting tree are detected and repaired. An AVL tree guarantees that the height difference of any node—defined as the height of the node's left subtree minus the height of the node's right subtree—is −1, 0, or 1.

As shown in Listing 6-9, each `BinaryNode` must store its `height` in the binary search tree. Whenever a node is inserted into the binary search tree, the height of the affected nodes must be computed *so an unbalanced tree node can be detected immediately*.

Listing 6-9. Structure of AVL binary node

```
class BinaryNode:
  def __init__(self, val):
    self.value = val                                        ❶
    self.left  = None
    self.right = None
    self.height = 0                                         ❷

  def height_difference(self):                              ❸
    left_height = self.left.height if self.left else -1     ❹
    right_height = self.right.height if self.right else -1
    return left_height - right_height                       ❺

  def compute_height(self):                                 ❻
    left_height = self.left.height if self.left else -1
    right_height = self.right.height if self.right else -1
    self.height = 1 + max(left_height, right_height)
```

❶ Structure of a `BinaryNode` is essentially the same as a binary search tree.

❷ Record the height for each `BinaryNode`.

---

4 Named after the inventors Adelson-Velsky and Landis.

❸ Helper function that computes the height difference between left and right sub-tree.

❹ Set left_height to −1 for nonexistent left subtree, or its proper height.

❺ Return height difference, which must be left_height subtracting right_height.

❻ Helper function that updates the height for a node *assuming that the* height *of its respective* left *and* right *subtrees (if they exist) have accurate* height *values.*

Listing 6-10 shows that the node returned by _insert() has its height properly computed.

Listing 6-10. Modify_insert() to compute height properly

```
def _insert(self, node, val):
  if node is None:
    return BinaryNode(val)              ❶

  if val <= node.value:
    node.left = self._insert(node.left, val)
  else:
    node.right = self._insert(node.right, val)

  node.compute_height()                 ❷
  return node
```

❶ For the base case, when a newly created leaf node is returned, its height is already 0 by default.

❷ When the recursive case completes, val has been inserted into either node.left or node.right. This means the height for node needs to be recomputed.

During the invocation of insert(27), a new leaf node for 27 is added to the binary search tree at the end of a sequence of recursive invocations, depicted in Figure 6-11. The final invocation to _insert() involves the base case where a new leaf node containing 27 is returned. This figure captures the brief moment when both the new leaf node (for 27) and the original leaf node (for 29) have a height of 0. With just one additional statement to compute the node's height at the end of _insert(), as the recursion unwinds, the height for each ancestor node (highlighted in Figure 6-11) are recomputed—note that these are the only nodes in the binary search tree whose heights need to be adjusted. The compute_height() function captures the logical definition for the height of a node, namely, that it is one greater than the larger of the heights of its children subtrees.

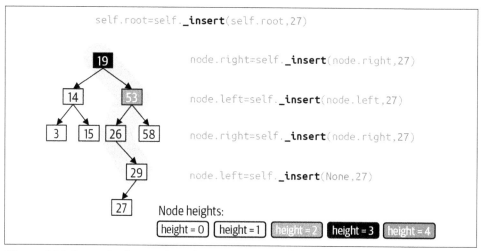

Figure 6-11. Recursive invocation when inserting a value

As the recursive invocations unwind, each ancestor node to 27 has its height recomputed. Because each node has an accurate `height` in the binary search tree, `_insert()` can detect whenever a node has become unbalanced—that is, *when the height of the left and right subtrees for that node differ by more than 1.*

In an AVL tree, the height difference for any node is -1, 0, or 1. The height difference is computed as the height of the node's left subtree minus the height of the node's right subtree. If a subtree doesn't exist, then use -1 for its height.

The node containing 26 leans to the right because its height difference is $-1 - 1 = -2$; the node containing 53 leans to the left because its height difference is $2 - 0 = 2$; lastly the root node leans to the right because its height difference is $1 - 3 = -2$. Once these nodes are identified, there needs to be a strategy to adjust the tree in some way to bring it back into balance. In the same way that the height is computed as the recursion unwinds, the `_insert()` function can immediately detect when the insertion of a new node has unbalanced the tree. It detects the imbalance as the recursion unwinds, which means the first unbalanced node detected is 26.

The designers of the AVL tree invented the concept of a *node rotation*, which is best described visually in Figure 6-12. The three nodes—containing the values 10, 30, and 50—are shaded to present their height. The root node, containing 50, has height h. The gray triangles are subtrees whose values conform to the binary search tree property. The left subtree of the node containing 10, for example, is labeled 10L, and it contains values that are all smaller than or equal to 10. All you need to know is that the height of this subtree (and the other three shaded subtrees, 10R, 30R, and 50R) is h – 3.

The tree leans to the left: its left subtree has a height of h – 1, while its right subtree has a smaller height of h – 3, meaning the height difference is +2. An AVL tree rebalances itself by detecting this imbalance and rotating nodes to reconfigure the tree, as shown on the right in Figure 6-12. After the rotation, the resulting binary search tree has a height of h – 1, and the node containing 30 has become the new root. This particular rotation is a *rotate right*, which you can visualize as placing your hand on the original node containing 30 and rotating your hand to the right, which "lifts up" the node containing 30 while "dropping down" the node containing 50.

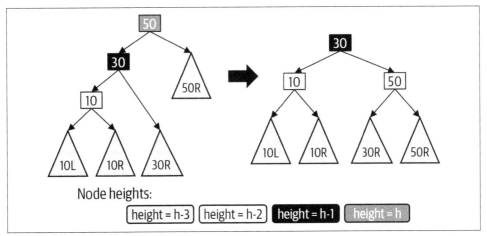

Figure 6-12. Rebalancing this binary search tree by rotating the root node to the right

There are four possible unbalanced scenarios in a binary search tree containing just three values, as shown in Figure 6-13. Scenario Left-left represents a simplified version of the example in Figure 6-12, which only needs a *rotate right* to balance the tree; similarly, scenario Right-right represents its mirror image, needing only a *rotate left* to bring the tree back into balance. These scenarios are named for the relative positioning of each descendant node to the root. These rotate operations result in a balanced tree whose root node contains 30, with a left child containing 10 and a right child containing 50.

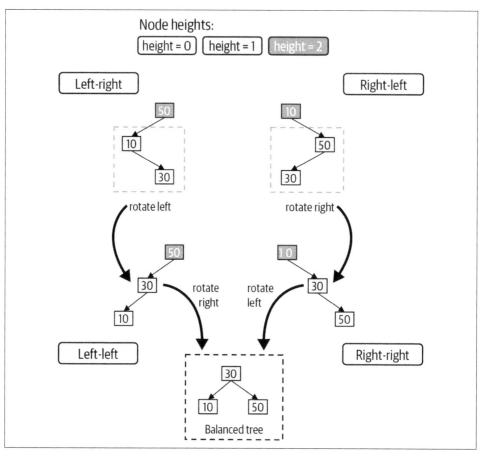

Figure 6-13. Four different node rotations

Scenario Left-right presents a more complicated unbalanced tree that can be rebalanced in two steps. First, *rotate left* the left subtree rooted at 10, which "drops down"
the 10 node and "lifts up" the 30 node, resulting in a tree that matches scenario Left-
left. Second, perform a *rotate right* to balance the tree. This two-step composite operation is called *rotate left-right*. Scenario Right-left represents the mirror image of
scenario Left-right, resulting in a *rotate right-left* operation that rebalances the tree.
The repository contains an optimized implementation for these composite
operations.

Two new helper functions resolve situations when a node is unbalanced to the left (or
to the right), as shown in Listing 6-11.

Listing 6-11. Helper functions that choose appropriate rotation strategy

```
def resolve_left_leaning(node):                    ❶
  if node.height_difference() == 2:
    if node.left.height_difference() >= 0:         ❷
      node = rotate_right(node)
    else:
      node = rotate_left_right(node)               ❸
  return node                                      ❼

def resolve_right_leaning(node):
  if node.height_difference() == -2:               ❹
    if node.right.height_difference() <= 0:        ❺
      node = rotate_left(node)
    else:
      node = rotate_right_left(node)               ❻
  return node                                      ❼
```

❶ A node leans to the left when height difference is +2.

❷ Detects the `rotate_right` case by confirming that node's `left` subtree is partially leaning left.

❸ Otherwise, node's `left` subtree is partially leaning right, meaning a `rotate_left_right` is in order.

❹ A node leans to the right when height difference is –2.

❺ Detects the `rotate_left` case by confirming that node's `right` subtree is partially leaning right.

❻ Otherwise, node's `right` subtree is partially leaning left, meaning a `rotate_right_left` is in order.

❼ Be sure to remember to return node of (potentially rebalanced) subtree.

The strategy is to immediately resolve an unbalanced node once this situation has been detected. The final implementation of _insert() is shown in Listing 6-12, which takes immediate advantage of these resolution helper functions. Adding a value to a left subtree of a node can never make that node right-leaning; similarly, adding a value to a right subtree of a node can never make that node left-leaning.

Listing 6-12. Rotating nodes when an unbalanced node is detected

```
def _insert(self, node, val):
  if node is None:
    return BinaryNode(val)

  if val <= node.value:
    node.left = self._insert(node.left, val)
    node = resolve_left_leaning(node)          ❶
  else:
    node.right = self._insert(node.right, val)
    node = resolve_right_leaning(node)         ❷

  node.compute_height()
  return node
```

❶ If left subtree is now left-leaning, resolve it.

❷ If right subtree is now right-leaning, resolve it.

The implementations for these rotation functions can be found in the code repository. Table 6-4 describes the rotate_left_right case, showing the code and the rebalanced tree. At the top, the new_root and the other affected nodes and subtrees are identified. Below, the tree is rebalanced, and, importantly, new heights are computed for child and node.

Pay attention to how rotate_left_right() returns the new root node for the balanced binary tree, since this unbalanced node could exist within a larger binary tree. There is no need to recompute the height of new_root since the calling function—in _insert() or _remove()—will do that. You can confirm visually that the resulting binary tree still conforms to the binary search tree property: for example, all of the values in 30L are greater than or equal to 10 and less than or equal to 30, so this subtree can be a right subtree of the node containing 10. A similar argument explains why the subtree labeled 30R can be a left subtree to the node containing 50.

Table 6-4. Implementation of rotate left-right

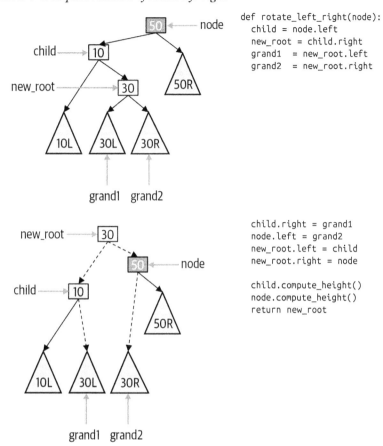

```
def rotate_left_right(node):
    child = node.left
    new_root = child.right
    grand1 = new_root.left
    grand2 = new_root.right
```

```
child.right = grand1
node.left = grand2
new_root.left = child
new_root.right = node

child.compute_height()
node.compute_height()
return new_root
```

The revised `_insert()` method now rebalances the binary search tree as needed. A similar change to `_remove()` and `_remove_min()` is straightforward with these helper functions, as shown in Listing 6-13. The code is modified to include four targeted interventions whenever the tree's structure is changed.

Listing 6-13. Updating _remove() to maintain AVL property

```
def _remove_min(self, node):
  if node.left is None: return node.right

  node.left = self._remove_min(node.left)
  node = resolve_right_leaning(node)          ❶
  node.compute_height()
  return node

def _remove(self, node, val):
  if node is None: return None

  if val < node.value:
    node.left = self._remove(node.left, val)
    node = resolve_right_leaning(node)        ❷
  elif val > node.value:
    node.right = self._remove(node.right, val)
    node = resolve_left_leaning(node)         ❸
  else:
    if node.left is None:  return node.right
    if node.right is None: return node.left

    original = node
    node = node.right
    while node.left:
      node = node.left

    node.right = self._remove_min(original.right)
    node.left = original.left
    node = resolve_left_leaning(node)         ❹

  node.compute_height()
  return node
```

❶ Removing the minimum value from a subtree rooted at node.left could make node right-leaning; rotate to rebalance as needed.

❷ Removing a value from the left subtree of node could make node right-leaning; rotate to rebalance as needed.

❸ Removing a value from the right subtree of node could make node left-leaning; rotate to rebalance as needed.

❹ After the minimum has been removed from the subtree returned to be node.right, node could be left-leaning; rotate to rebalance as needed.

The AVL implementation now properly rebalances whenever a new value is inserted into the tree or a value is removed. Each rebalancing contains a fixed number of operations and executes in O(1) constant time. Because an AVL tree is still a binary search tree, the search and traversal functions do not need to change.

Analyzing Performance of Self-Balancing Trees

The `compute_height()` helper function and the different node rotation methods all perform in a constant amount of time—there are no further recursive calls or loops in any of these functions. These tree maintenance functions are invoked *only* when a node is detected to be unbalanced. It turns out that when inserting a value into an AVL tree, there will never be more than one node rotation required. When removing a value, it is theoretically possible there will be multiple node rotations (see the challenge exercise at the end of this chapter that investigates this behavior). In the *worst case*, there will never be more `log` (N) rotations, which means that the runtime performance for search, insert, and remove are all O(`log` N).

Using the information from this chapter, you are now prepared to further investigate any recursive data structures. To close this chapter, I now reconsider the symbol table and priority queue data types to consider whether a binary tree can provide a more efficient implementation.

Using Binary Tree as (key, value) Symbol Table

The same binary search tree structure can be used to implement the symbol table data type introduced in Chapter 3, as shown in Figure 6-14.

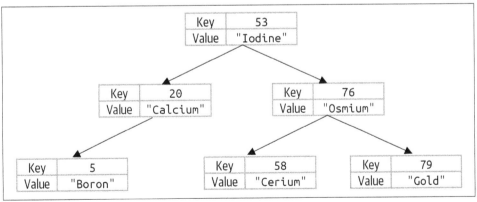

Figure 6-14. Binary search tree as symbol table: keys are atomic numbers; values are element names

To do this, you have to modify the `BinaryNode` structure to store both `key` and `value`, as shown in Listing 6-14.

Listing 6-14. Updated `BinaryNode` when using binary tree to store symbol table

```
class BinaryNode:
  def __init__(self, k, v):
    self.key = k          ❶
    self.value = v        ❷
    self.left = None
    self.right = None
    self.height = 0
```

❶ The key is used to navigate the binary search tree.

❷ The `value` contains arbitrary data that is irrelevant to the operation of the binary search tree.

`BinaryTree` now needs `put(k,v)` and `get(k)` functions, instead of `insert()` and `__contains()__`, to support the expected interface for a symbol table. The changes are minimal, and only incidental changes are required, so the code is not reproduced here; find it in the associated code repository (*https://oreil.ly/fUosk*). When navigating through the binary search tree, that is, whether to go `left` or `right`, the decision is based on `node.key`.

Using a binary search tree provides the added benefit that the keys can be retrieved from the symbol table *in ascending order* using the `__iter()__` traversal function.

Chapter 3 described how open addressing and separate chaining can implement the symbol table data type. It's worth comparing the runtime performance of a binary search tree against the results of open addressing and separate chaining hashtables, as shown in Table 3-4. This trial inserts N = 321,129 words from the English dictionary into a symbol table. What is the smallest height for a binary tree that stores all of these words? Recall that this height is computed by the formula $\log(N + 1) - 1$, which is 17.293. After inserting all of these words *in ascending order* from the English dictionary, the height of the resulting AVL binary search tree is 18, which further demonstrates the efficiency of AVL trees in storing information.

The hashtable implementations from Chapter 3 *significantly outperform* binary search trees, as shown in Table 6-5. If you ever need the keys for a symbol table in ascending order, then I recommend retrieving the keys from the symbol table and then sorting them separately.

Table 6-5. Comparing AVL symbol table implementation with hashtables from Chapter 3 (time in seconds)

Type	Open addressing	Separate chaining	AVL trees
Build time	0.54	0.38	5.00
Access time	0.13	0.13	0.58

Using the Binary Tree as a Priority Queue

Given that the heap data structure described in Chapter 4 is based on a binary tree structure, it is only natural to compare the runtime performance of a priority queue implemented using an AVL binary search tree where the `priority` is used to navigate the structure of the binary search tree, as shown in Figure 6-15.

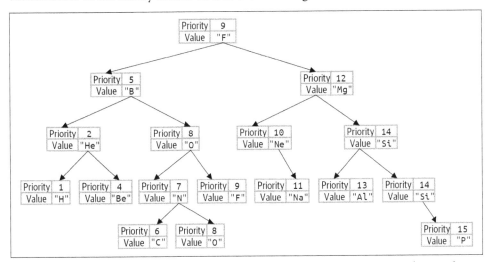

Figure 6-15. Binary search tree as priority queue: priorities are atomic numbers; values are element symbols

There are two benefits to using a binary search tree to implement a priority queue:

- An array-based heap must create storage for a fixed number of values in advance. Using a binary search tree, the structure can grow to be as large as needed.

- In the heap structure, there is no way to provide an iterator of the entries in the priority queue *in priority order* without dequeuing the values.[5] With a binary search tree structure, this capability now exists using the traversal logic.

To get started, `BinaryNode` now stores both `value` and `priority`, as shown in Listing 6-15. The `priority` field will be used to navigate the binary tree for the search, insert, and remove operations.

Listing 6-15. Updated BinaryNode when using binary tree to store priority queue

```
class BinaryNode:
  def __init__(self, v, p):
    self.value = v           ❶
    self.priority = p        ❷
    self.left  = None
    self.right = None
    self.height = 0
```

❶ The `value` contains arbitrary data that is irrelevant to the operation of the binary search tree.

❷ The `priority` is used to navigate the binary search tree.

In a max binary heap, the entry with highest priority is in `storage[1]`, and it can be located in $O(1)$ constant time. This is not the case when using a binary search tree to store a priority queue. The `BinaryNode` with highest priority is the right-most node in the binary tree. To locate this value requires $O(\log N)$ runtime performance, if the underlying binary tree is balanced using the techniques described in this chapter.

The biggest change, however, is that when a priority queue uses a binary search tree for storage, the only value to remove *is the one with highest priority*; this means the general purpose `remove()` function is not needed. In its place, a `_remove_max()` helper function is added to PQ, as shown in Listing 6-16. The other helper functions are part of the standard priority queue interface. Note that the count, N, of pairs is stored and managed by the PQ class.

---

5 See challenge exercise at end of this chapter showing how to do this.

Listing 6-16. PQ class provides enqueue() and dequeue() functions

```
class PQ:
  def __init__(self):
    self.tree = BinaryTree()                                    ❶
    self.N = 0

  def __len__(self):
    return self.N

  def is_empty(self):
    return self.N == 0

  def is_full(self):
    return False

  def enqueue(self, v, p):
    self.tree.insert(v, p)                                      ❷
    self.N += 1

  def _remove_max(self, node):                                  ❸
    if node.right is None:
      return (node.value, node.left)                            ❹

    (value, node.right) = self._remove_max(node.right)          ❺
    node = resolve_left_leaning(node)                           ❻
    node.compute_height()                                       ❼
    return (value, node)

  def dequeue(self):                                            ❽
    (value, self.tree.root) = self._remove_max(self.tree.root)
    self.N -= 1
    return value                                                ❾
```

❶   Use a balanced binary search tree for storage.

❷   To enqueue a (v, p) pair, insert that pair into the binary search tree and increment
    N count.

❸   The _remove_max() helper method both removes the node with maximum prior-
    ity from the subtree rooted at node *and* returns its value and the node of the
    resulting subtree as a tuple.

❹   Base case: with no right subtree, this node has maximum priority; return both the
    value in the node being deleted and the left subtree that will eventually take its
    place.

❺   Recursive case: retrieve removed value and root of updated subtree.

❻ If node is out of balance (it could now lean left), fix with rotations.

❼ Compute node height before returning it along with value that was removed.

❽ The dequeue() method removes node with maximum priority from the binary search tree and returns its value.

❾ After decrementing count, N, return the value that had been associated with highest priority.

The runtime performance of this priority queue implementation still offers O(log N) behavior, although in absolute terms it is twice as slow as the heap-based priority queue implementation from Chapter 4. The reason is that maintaining the AVL binary search tree structure is more work than actually needed for a priority queue. Still, if you need the ability to iterate over its (value, priority) pairs in the order in which they would be removed, this is an efficient alternative.

Summary

Binary trees are a dynamic, *recursive* data structure that organizes its values into *left* and *right* substructures, offering the potential to evenly subdivide a collection of N values into two structures, each containing (more or less) N/2 values. Binary trees form the basis for countless other recursive data structures that lead to efficient implementations, including:

- Red-black trees, which offer a more efficient approach for balancing binary search trees, although the implementation is more complicated than with AVL trees.
- B-trees and B+ trees, used for databases and file systems.
- R-trees and R* trees, used for spatial information processing.
- *k*-d trees, Quadtrees, and Octrees for space-partitioning structures.

To summarize:

- Because trees are recursive data structures, it is natural to write recursive functions to manipulate their structure.
- The most common technique for traversing a binary search tree is *inorder traversal*, which returns all values in ascending order. The Expression recursive structure includes a *postorder* traversal function that produces the values in postfix order, which conforms to the postfix notation, which is used by some handheld calculators.

- Binary search trees must rebalance their structure to ensure they can achieve O(log N) runtime performance for its key operations. The AVL technique is able to balance a tree by enforcing the AVL property, that the height difference of any node is –1, 0, or 1. To make this work efficiently, each binary node also stores its *height* in the tree.
- A priority queue can be implemented using a balanced binary search tree to store the (`value`, `priority`) pairs, using `priority` when comparing nodes. One benefit of this structure is you can use inorder traversal to return the pairs stored by the priority queue in priority order, *without affecting the structure of the priority queue.*
- A symbol table can be implemented using a balanced binary search tree by enforcing the restriction that each key is unique in the binary search tree. However, the performance will not be as efficient as the hashtable implementations described in Chapter 3.

Challenge Exercises

1. Write a recursive `count(n, target)` function that returns the number of times that `target` exists within the linked list whose first node is `n`.

2. Sketch the structure of a binary search tree with N nodes that requires O(N) time to find the two largest values. Next, sketch the structure of a binary search tree with N nodes that requires O(1) time to find the two largest values.

3. What if you wanted to find the kth smallest key in a binary search tree? An inefficient approach would be to traverse the entire tree until k nodes have been visited. Instead, add a function, `select(k)`, to `BinaryTree` that returns the kth smallest key for k from 0 to N – 1. For an efficient implementation, you will need to augment the `BinaryNode` class to store an additional field, N, that records the number of nodes in the subtree rooted at that node (including that node itself). A leaf has an N value of 1, for example.

    Also add the companion method, `rank(key)`, to `BinaryTree` that returns an integer from 0 to N – 1 that reflects the rank of `key` in sorted order (or in other words, the number of keys in the tree strictly less-than `key`).

4. Given the values [3,14,15,19,26,53,58], there are 7! = 5,040 ways to insert these seven values into an empty binary search tree. Compute the number of different ways that the resulting tree is perfectly balanced, with a height of 2, such as the binary search tree shown in Figure 6-3.

    Can you generalize your result to an arbitrary collection of 2^{k-1} values and present a recursive formula $c(k)$ that computes this for any k?

5. Write a contains(val) method for BinaryTree that invokes a recursive contains(val) method in BinaryNode.

6. As described in this chapter, AVL trees are self-balancing. For a given N, can you compute the maximum height of an AVL tree containing N values, since they cannot all be so perfectly compact as a complete binary tree? Generate 10,000 random AVL trees of size N, and record the maximum observed height for each N.

Create a table that records whenever this maximum observed height increases. Predict the values of N such that an AVL tree of N nodes can have a tree height that is one greater than any AVL tree with N − 1 nodes.

7. Complete the SpeakingBinaryTree in Listing 6-17 whose insert(val) operation produces English descriptions of the actions as they are performed. Table 6-2 contains the desired output for each of the corresponding operations. This recursive operation is different from others in this chapter because it processes "top-down," whereas most recursive functions process "bottom-up" from base cases.

Listing 6-17. Enhance the _insert() method to return description of what happened

```
class BinaryNode:
  def __init__(self, val):
    self.value = val
    self.left  = None
    self.right = None

class SpeakingBinaryTree:
  def __init__(self):
    self.root = None

  def insert(self, val):
    (self.root,explanation) = self._insert(self.root, val,
        'To insert `{}`, '.format(val))
    return explanation

  def _insert(self, node, val, sofar):
    """
    Return (node,explanation) resulting from inserting val into subtree
    rooted at node.
    """
```

Modify the _insert() function to return a tuple (node, explanation), where node is the resulting node and explanation contains the growing explanation for the actions.

8. Write a method `check_avl_property(n)` that validates the subtree rooted at n to ensure that (a) that each descendant node's computed `height` is correct, and (b) that each descendant node satisfies the AVL tree property.

9. Write a `tree_structure(n)` function that produces a string with parentheses in *prefix order* to capture the structure of the binary tree rooted at n. In prefix order, the value for the node is printed first, before the left representation and right representation. This string should use commas and parentheses to separate information, so it can be parsed later. For the complete binary tree in Figure 6-3, the resulting string should be `'(19,(14,(3,,),(15,,)),(53,(26,,),(58,,)))'`, while for the binary tree in the left side of Figure 6-5, the resulting string should be `'(5,(4,(2,(1,,),(3,,)),),(6,,(7,,)))'`.

Write the companion `recreate_tree(expr)` function that takes in an `expr` tree structure string using parentheses and returns the root node of a binary tree.

10. If you count *rotate left-right* and *rotate right-left* as single rotations (in addition to *rotate left* and *rotate right*), then you will never need more than a single rotation when inserting a value into an AVL binary search tree. However, when removing a value from an AVL tree, you might need multiple rotations.

What is the smallest AVL binary search tree that requires multiple node rotations when a single value is removed? Such a tree would have to have at least four nodes. To answer this question, you will have to instrument the rotation methods to increase a count for the number of times a rotation is used. Also, use the results of `tree_structure()` in the previous exercise to record the tree so you can recover its structure *after* you have detected the multiple rotations. Write a function that (a) generates 10,000 random AVL trees containing between 4 and 40 nodes, and (b) for each of these trees, select one of its random values to remove. You should be able to compute the size of the AVL trees that require up to three rotations for a remove request. As a hint, you should be able to generate an AVL tree with 4 nodes that requires 1 rotation for a given remove request, and an AVL tree with 12 nodes that requires 2 rotations for a given remove request. What is the smallest AVL tree that you can find which requires 3 rotations for a given remove request?

11. A complete binary tree with $N = 2^k - 1$ nodes is the most compact representation for storing N nodes. This question asks what is the "least compact" AVL tree you can construct. A *Fibonacci tree* is an AVL tree such that in every node, the height of its left subtree is bigger (by just 1) than the height of its right subtree. Think of this as an AVL tree that is one insert away from rebalancing. Write a recursive function, `fibonacci_avl(N)`, for $N > 0$ that returns a `BinaryNode` representing the root of a Fibonacci tree. It is simpler to do this without involving any `Binary Tree` objects. The root node returned contains the value F_N. For example,

`fibonacci_avl(6)` would return the root node for the binary tree depicted in Figure 6-16.

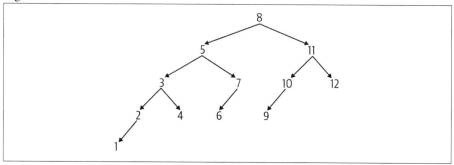

Figure 6-16. A Fibonacci tree with twelve nodes

## CHAPTER 7

Graphs: Only Connect!

In this chapter, you will learn:

- The *stack* abstract data type.
- The *indexed min priority queue* data type, which is the final data type included in this book.
- How to model a *graph* using nodes and edges. In a directed graph, the edges have an orientation. In a weighted graph, edges have an associated numeric value.
- How Depth First Search uses a stack to organize searching in a graph.
- How Breadth First Search uses a queue to search through a graph. If a path exists between a source node and a target node, Breadth First Search will return the shortest path that exists.
- How to detect whether a directed graph contains a *cycle*: a sequence of edges starting from, and ending with, a specific node.
- How to use Topological Sort in a directed graph to produce a linear ordering of nodes compatible with all dependencies in the directed graph.
- How to determine the shortest accumulated path in a weighted graph from one node to all other nodes.
- How to determine the shortest accumulated path in a weighted graph between any two nodes.

Graphs Efficiently Store Useful Information

I've covered algorithms for solving common problems in information systems regarding storing and processing data. These algorithms can solve countless real-world problems if only we could properly model these problems. Here are three such problems that I will solve by using *graphs*:

- A maze consists of rooms with doorways leading to other rooms. Find the shortest path from an entrance to the exit.

- A project is defined by a collection of tasks, but some tasks require other tasks to complete before they can begin. Assemble a linear schedule that describes the order in which the tasks can be performed to complete the project.

- A map contains a collection of highway segments, including their length in miles. Find the shortest traveling distance between any two locations in the map.

Each of these problems can be modeled effectively using *graphs*, a fundamental concept studied by mathematicians for centuries. Modeling the relationships *between* data is often as important as the data values themselves. A *graph* models information as *nodes* connected by *edges*. Any number of edges, e = (u, v), can exist to represent some relationship between nodes *u* and *v*. As you can see in Figure 7-1, graphs can model concepts from a variety of application domains. An *undirected graph* can model the structural relationship between the carbon and hydrogen atoms in the propane molecule. A mobile app can provide driving directions in New York City by representing the orientation of one-way streets as a *directed graph*. A driver's road atlas can represent the driving distances between New England state capitals as a *weighted graph*. With a bit of computation, you can see that the shortest driving distance from Hartford, Connecticut, to Bangor, Maine, is 278 miles.

A *graph* is a data type that contains a collection of N distinct nodes, each with a unique label to identify the node.[1] You can add an *edge* to a graph to connect two different nodes, *u* and *v*, with each other. An edge is represented as (u, v), and *u* and *v* are called its *endpoints*. Each edge (u, v) joins *u* and *v* together so *u* is adjacent to *v* (and, vice versa, *v* is adjacent to *u*).

---

1 A node is often called a vertex, but for this chapter I use the term *node* to be consistent with networkx.

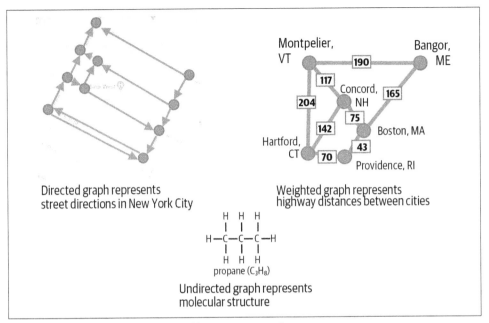

Figure 7-1. Modeling different problems using graphs

The graph in Figure 7-2 has 12 unique nodes and 12 edges. Imagine that each node is an island, and the edges are bridges connecting the islands. A traveler can walk from island B2 to island C2 or from island C2 to island B2; however, there is no way for a traveler to *walk directly* from island B2 to B3. Instead, the traveler can cross the bridge from island B2 to island C2, then cross the bridge from island C2 to island C3, and finally cross the bridge from island C3 to B3. Based on this representation of the islands and its bridges, the traveler can find a sequence of bridges to travel between any of the "B" and "C" islands, but despite the bridges connecting the "A" islands, there is no way to travel from an "A" island to a "B" island.

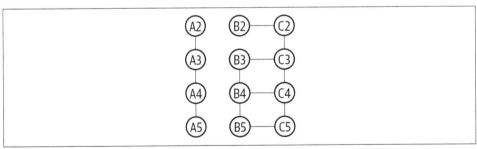

Figure 7-2. An undirected graph with 12 nodes and 12 edges

Given a graph containing nodes and edges, a common problem is to compute a path from a *source* node (such as the node representing island B2) to a *target* node (such as the node representing the island B3) using only the edges contained in the graph.

A *path* is formed by a sequence of edges starting at a source and terminating at a target. Naturally, every edge forms a path between its two endpoint nodes, but what about two nodes that do not have an edge to join them together? In Figure 7-2 there is a path from node B2 to B3 by following the sequence of edges (B2, C2), (C2, C3), and then (C3, B3).

In a path, each successive edge must start at the node in which the prior edge terminated. A path can also be represented by listing the sequence of nodes encountered along the way, such as [B2, C2, C3, B3]. Another longer path from B2 to B3 is [B2, C2, C3, C4, B4, B3]. A *cycle* is a path that starts and ends at the same node, such as [C4, C5, B5, B4, C4]. A node, *v*, is *reachable* from another node, *u*, if there is a path of edges in the graph from *u* to *v*. In some graphs, there may be no path between two nodes—for example, there is no path from A2 to B2 in the graph from Figure 7-2. When this happens, the graph is considered to be *disconnected*. In a *connected* graph, it is possible to compute a path between any two of its nodes.

In this chapter, I will present three types of graphs:

Undirected graph
    A graph that contains edges (*u*, *v*) that connect two nodes such that *u* is adjacent to *v*, and *v* is adjacent to *u*. This is like a bridge that can be traveled in either direction.

Directed graph
    A graph that contains edges (*u*, *v*), where each edge has a fixed orientation. When the edge (*u*, *v*) is in the graph, *v* is adjacent to *u*, but the opposite is not true. This is like a one-way bridge that a traveler can use to walk from island *u* to island *v* (but not the other direction).

Weighted graph
    A graph that contains edges (*u*, *v*, *weight*), where *weight* is a numeric value associated with the edge (note: the underlying graph can either be directed or undirected). This weight represents an aspect about the relationship between *u* and *v*; for example, weight could represent the physical distance in miles between the locations modeled by the nodes *u* and *v*.

All graphs in this chapter are *simple* graphs, which means each edge is unique (i.e., there cannot be multiple edges between the same pair of nodes), and there are no self-loops where an edge connects a node with itself. Either a graph contains all undirected edges or it has all directed edges. Similarly, either a graph has all weighted edges or none of the edges have associated weights.

For the algorithms in this chapter, you need a graph data type that can provide the following functionalities:

- Return the number of nodes, N, and the number of edges, E, in the graph.
- Generate the collection of nodes and edges.
- Generate the adjacent nodes or edges for a given node.
- Add a node or edge to a graph.
- Remove a node or edge from a graph—this functionality is not essential for the algorithms presented in this chapter, but I include it for completeness.

Python has no built-in data structure that provides this functionality. Instead of implementing code from scratch, you need to install the NetworkX (*https:// networkx.org*) open source library to create and manipulate graphs. Doing so ensures you do not waste time reinventing the wheel, plus it gives you access to an impressive number of graph algorithms already implemented by networkx. In addition, net workx seamlessly integrates with other Python libraries to visualize graphs. The program in Listing 7-1 constructs the graph shown in Figure 7-2.

Listing 7-1. A program that builds the graph in Figure 7-2

```
import networkx as nx
G = nx.Graph()                                          ❶
G.add_node('A2')                                        ❷
G.add_nodes_from(['A3', 'A4', 'A5'])                    ❸

G.add_edge('A2', 'A3')                                  ❹
G.add_edges_from([('A3', 'A4'), ('A4', 'A5')])          ❺

for i in range(2, 6):
  G.add_edge('B{}'.format(i), 'C{}'.format(i))          ❻
  if 2 < i < 5:
    G.add_edge('B{}'.format(i), 'B{}'.format(i+1))
  if i < 5:
    G.add_edge('C{}'.format(i), 'C{}'.format(i+1))

>>> print(G.number_of_nodes(), 'nodes.')               ❼
>>> print(G.number_of_edges(), 'edges.')
>>> print('adjacent nodes to C3:', list(G['C3']))      ❽
>>> print('edges adjacent to C3:', list(G.edges('C3'))) ❾
12 nodes.
12 edges.
adjacent nodes to C3: ['C2', 'B3', 'C4']
edges adjacent to C3: [('C3', 'C2'), ('C3', 'B3'), ('C3', 'C4')]
```

❶ nx.Graph() constructs a new undirected graph.

❷   A node can be any hashable Python object except None. Strings are a good choice.

❸   Add multiple nodes from a list using add_nodes_from().

❹   Add an edge between two nodes, *u* and *v*, with add_edge(u, v).

❺   Add multiple edges from a list using add_edges_from().

❻   If an edge is added to a graph *before its nodes are*, the corresponding nodes are automatically added to the graph.

❼   A graph can report its number of nodes and edges.

❽   Find the nodes adjacent to v by using the G[v] lookup capability.

❾   Find the edges adjacent to v by using the G.edges(v) function.

You might wonder about the order in which adjacent nodes (or edges) are returned when requested. In the subsequent code, when adjacent edges or nodes are requested, you cannot expect that they will be returned in a specific order.

Using Depth First Search to Solve a Maze

Given a rectangular maze as shown in Figure 7-3, how would you write a program that solves it? The entrance to this 3 x 5 maze consisting of 15 cells is at the top, and the desired exit is at the bottom. To move through the maze, you can only move horizontally or vertically between rooms that are not blocked by walls. The first step is to model the maze using an undirected graph consisting of 15 nodes, where each node, labeled (row, column), models a cell in the maze. The *source* of the maze, for example, is labeled (0, 2), and the *target* is labeled (2, 2). The second step is to add an edge between two nodes (*u*, *v*) if their corresponding cells in the maze *do not have a wall between them*. The resulting graph is shown overlayed with the maze so you can see the one-to-one correspondence between a cell in the maze and a node in the graph.

Finding a path between the source node (0, 2) and target node (2, 2) is equivalent to finding a solution to the original rectangular maze. I will show you a technique for solving any such maze, regardless of its size. If you try to solve this maze on your own, you will explore different paths, discarding those that lead to "dead ends," until you eventually find a solution. Though you might not realize it, you have a significant advantage because *you can see the whole maze at a glance* and can make decisions on which paths to explore based on your own sense of *how close you are to the final target*. Imagine, instead, that you are stuck inside the maze,[2] and you can only see the cells that directly connect to the cell in which you stand—these restrictions completely change your approach.

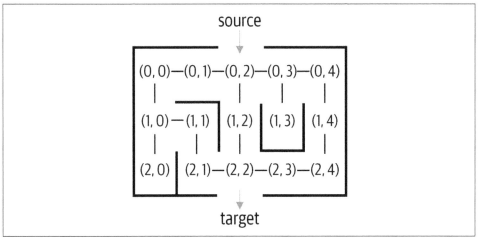

Figure 7-3. A graph modeling a rectangular maze

Let's develop a strategy to solve a maze using its corresponding undirected graph, as seen in Figure 7-4. These mazes are randomly generated, which is an interesting exercise all on its own.[3]

Starting from the source (0, 2), you see three adjacent nodes that are connected by edges; you arbitrarily head east to (0, 3) but remember (0, 1) and (1, 2) as potential points to explore. Node (0, 3) has three adjacent nodes, but *you remember that you came from* (0, 2), and you don't repeat where you've already been, so you arbitrarily head south to (1, 3) but remember (0, 4) as a potential point to explore. You have just followed the highlighted path shown in Figure 7-4 and have reached a dead end.

You know (1, 3) is a dead end because *there is no adjacent node you have not already seen*. What should you do? In Figure 7-4, I circled those nodes you came across but didn't explore: perhaps your search will be more fruitful if *you backtrack to one of these previous nodes and continue searching from there.*

---

2 Which can happen in real-life corn mazes! *Cool Patch Pumpkins* in Dixon, California, is the largest corn maze in the world, measuring a total 63 acres. It takes several hours to complete.

3 See the ch07.maze program for details.

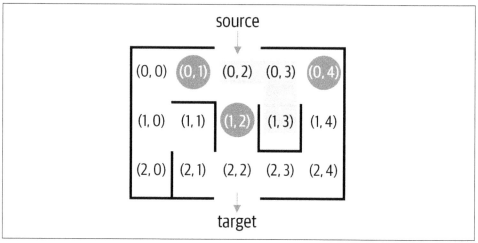

Figure 7-4. Hitting a dead end while exploring a maze

Here is a rough sketch of the activity of a graph-searching algorithm that explores a graph starting from a designated *source* node:

- Mark each node that you visit.
- Find the adjacent nodes to your current node that have not yet been marked as visited, and *arbitrarily select one of these* to explore.
- Go back to the last unmarked node you remembered when you hit a dead end.
- Continue exploring until all reachable nodes are marked.

Once the search algorithm completes, it should be possible to reconstruct the path from the source node to any node in the graph. To make this possible, the search algorithm must return a structure that contains enough information to support this capability. A common solution is to return a node_from[] structure, where node_from[v] either is None, if v is not reachable from the source node, or it is u, that is, *the prior node discovered before exploring* v. In Figure 7-4, node_from[(1, 3)] is the node (0, 3).

This sketch of an algorithm cannot be performed within a simple while loop, like when searching a linked list for a value. Instead, I now show how to use the *stack* abstract data type to maintain the search state while exploring the graph.

If you have ever eaten a meal at a cafeteria, you've no doubt grabbed the top tray from a stack of trays. The stack data type represents the behavior of such a stack of trays. A stack has a push(value) operation that adds value to become the newest value at the *top* of the stack, and pop() removes the value at the top of the stack. Another way to describe this experience is "Last in, first out" (LIFO), which is shorthand for "Last

[one] in [the stack is the] first [one taken] out [of the stack]." If you push three values, 1, 2, and 3, onto a stack, they will pop off the stack in order 3, 2, and finally 1.

Using the Node linked list data structure from Listing 6-1, the Stack implementation in Listing 7-2 has a push() operation to prepend a value to the front of a linked list. The pop() method removes and returns the first value in a linked list; as you can see, this provides the behavior for a stack. In Stack, the push() and pop() operations perform in constant time, independent of the total number of values in the stack.

Listing 7-2. Linked list implementation of Stack data type

```
class Stack:
  def __init__(self):
    self.top = None                            ❶

  def is_empty(self):
    return self.top is None                    ❷

  def push(self, val):
    self.top = Node(val, self.top)             ❸

  def pop(self):
    if self.is_empty():                        ❹
      raise RuntimeError('Stack is empty')

    val = self.top.value                       ❺
    self.top = self.top.next                   ❻
    return val
```

❶ Initially, top is None, reflecting an empty Stack.

❷ A Stack is empty if top is None.

❸ Ensures new Node is the first one in linked list, with existing linked list becoming the rest.

❹ An empty Stack causes a RuntimeError.

❺ Extract the newest value from top of stack to be returned.

❻ Reset Stack so the next Node is now on top (if None, then Stack becomes empty).

The Depth First Search algorithm uses a stack to keep track of marked nodes it will explore in the future. Listing 7-3 contains a stack-based implementation for Depth First Search. The search strategy is called *depth first* because it constantly tries to advance forward, always expecting that the solution is just one step away.

It starts at a source node, src, which is marked as having been visited (i.e., setting marked[src] to True) and is then pushed onto the stack for further processing. Each time through the while loop, the stack contains nodes that have already been visited and marked: they are popped from the stack, one at a time, and unmarked adjacent neighbors are marked and added to the stack for further processing.

Listing 7-3. Depth First Search of graph from designated source node, src

```
def dfs_search(G, src):              ❶
  marked = {}                        ❷
  node_from = {}                     ❸

  stack = Stack()
  marked[src] = True                 ❹
  stack.push(src)

  while not stack.is_empty():        ❺
    v = stack.pop()
    for w in G[v]:
      if not w in marked:
        node_from[w] = v             ❻
        marked[w] = True             ❼
        stack.push(w)

  return node_from                   ❽
```

❶ Conduct a Depth First Search over graph, G, starting from source node, src.

❷ The marked dictionary records nodes that have already been visited.

❸ Record how search got to each node: node_from[w] is the prior node working backward to src.

❹ Mark and place src node into Stack to start the search. The top node in the Stack represents the next node to explore.

❺ If the Depth First Search has not yet completed, v is the next node to explore.

❻ For each unmarked node, w, adjacent to v, remember that to get to w, the search came from v.

❼ Push w onto the top of the stack and mark it so it won't be visited again.

❽ Return the *structure of the search* that records for each node, v, the prior node from a search initiated at src.

Figure 7-5 visualizes the execution of Depth First Search, showing the updated state of the stack each time through the `while` loop. The highlighted node at the top of the stack is the current cell being explored; other nodes in the stack represent nodes that will *eventually be processed in the future*. You might wonder how Depth First Search avoids getting stuck wandering around aimlessly, forever. Each time a node is pushed onto the stack, it is marked, which means it will never be pushed onto the stack again. The `for` loop over w will find all unmarked nodes adjacent to v, yet to explore: it will mark each w and push w onto the stack for further exploration.

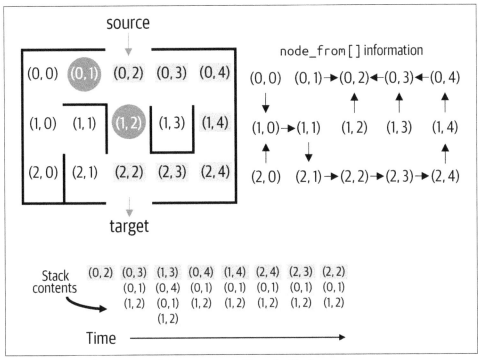

Figure 7-5. Depth First Search locates target if reachable from source

The search finds its first dead end, at (1, 3), but quickly recovers, popping off node (0, 4) to resume its search. The search is paused in Figure 7-5 to show the state once the target is found, but the search continues until all nodes in the graph reachable from src are explored and the stack becomes empty.

The stack will ultimately become empty, since (a) there are a finite number of nodes in the graph, and (b) unmarked nodes are first marked before being pushed onto the stack. Since a node can never become "unmarked," eventually each node that is reachable from src will be pushed exactly once onto stack and subsequently removed within the `while` loop.

The resulting depth-first *search tree* is shown on the right in Figure 7-5 as contained in the node_from[] structure. This structure can be called a *tree* because there are no cycles among the arrows. It encodes information that can be used to recover the path from (0, 2) to any reachable node in the graph by working *backward*. For example, node_from[(0, 0)] = (1, 0), which means the second-to-last node on the path from (0, 2) to (0, 0) was (1, 0).

The computed six-move solution is not the shortest possible path to the target. Depth First Search offers no guarantee regarding the length of the discovered path, but it will eventually find a path to every reachable node from a designated source node. Given the computed node_from[] structure resulting from a Depth First Search initiated at src, the path_to() function in Listing 7-4 computes the sequence of nodes from src to any target reachable from src. Each node_from[v] records the prior node encountered in a search from src.

Listing 7-4. Recovering actual path from node_from[]

```
def path_to(node_from, src, target):    ❶
    if not target in node_from:
        raise ValueError('Unreachable')    ❼

    path = []
    v = target                           ❷
    while v != src:
        path.append(v)                   ❸
        v = node_from[v]                 ❹

    path.append(src)                     ❺
    path.reverse()                       ❻
    return path
```

❶ node_from structure is needed to recover path from src to any target.

❷ To recover the path, set v to target node.

❸ As long as v is not src, append v to path, a backward list of nodes found on path from src to target.

❹ Continue backward by setting v to the prior node recorded by node_from[v].

❺ Once src is encountered, the while loop terminates, so src must be appended to complete the backward path.

❻ Return the reverse of path to produce the proper ordering from src to target.

---

❼ If node_from[] doesn't contain target, then it is not reachable from src.

The path_to() function computes the sequence of nodes in reverse order from tar get backward until src is encountered; it then simply reverses the order of the discovered nodes to produce a solution in the proper order. If you try to recover a path to an unreachable node, then path_to() will raise a ValueError.

Depth First Search repeatedly heads off in arbitrary directions with the expectation that it is just one node away from the destination. Now let's look at a more methodical search strategy.

Breadth First Search Offers Different Searching Strategy

Breadth First Search explores nodes *in order of their distance from the source.* Using the same maze from Figure 7-3, Figure 7-6 identifies each cell in the graph by *its shortest distance* from the source. As you can see, it finds a path through the maze just three cells long. In fact, Breadth First Search will always find the shortest path in a graph in terms of the number of edges visited.

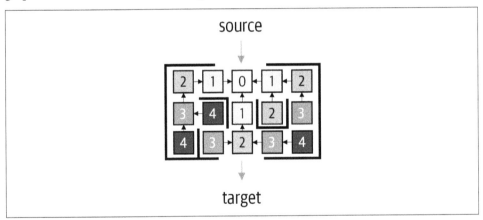

Figure 7-6. Breadth First Search locates the shortest path to target, if reachable from source

To provide some intuition behind Breadth First Search, observe in Figure 7-6 that from the source of the maze, there are three nodes that are just one step away—any one of these could lead to the shortest distance solution to the target, but, of course, you can't know which one. Instead of picking just one of these nodes to explore, Breadth First Search takes each one and advances one step further until it finds those nodes that are two steps away. With this methodical approach, it will not make any rash decisions while exploring the graph.

Instead of being optimistic like Depth First Search, Breadth First Search explores each of these nodes *in order* until all nodes that are one step away from the source have been visited; this results in four different nodes (labeled with 2 in Figure 7-6) that are two steps away from the source. In similar fashion, each of these nodes will be explored *in order* until all nodes that are two steps away from the source have been visited, resulting in four nodes that are three steps away from the source. This process continues until every node in the graph reachable from the source is visited.

Breadth First Search needs a structure to keep track of the nodes because it has to make sure that a node with distance d + 1 is not explored before all nodes with distance d are visited. The queue data type covered in Chapter 4 will process nodes in this order, because it enforces a "first in, first out" (FIFO) policy for adding and removing values. The code in Listing 7-5 is nearly identical to the code for Depth First Search, with the exception that a queue stores the *active search space*, that is, the nodes actively being explored.

Listing 7-5. Breadth First Search of graph from designated source node

```
def bfs_search(G, src):          ❶
  marked = {}                    ❷
  node_from = {}                 ❸

  q = Queue()
  marked[src] = True             ❹
  q.enqueue(src)

  while not q.is_empty():        ❺
    v = q.dequeue()
    for w in G[v]:
      if not w in marked:
        node_from[w] = v         ❻
        marked[w] = True         ❼
        q.enqueue(w)

  return node_from               ❽
```

❶ Conduct a Breadth First Search over graph, G, starting from source node, src.

❷ The marked dictionary records nodes that have already been visited.

❸ Record how search got to each node: node_from[w] is the prior node working backward to src.

❹ Mark and place src node into Queue to start the search. The first node in the Queue represents the next node to explore.

❺   If the Breadth First Search has not yet completed, v is the next node to explore.

❻   For each unmarked node, w, adjacent to v, remember that to get to w, the search came from v.

❼   Place w as last node to explore at the end of the queue, and mark it so it isn't visited multiple times.

❽   Return the *structure of the search* that records for each node, v, the prior node from a search initiated at src.

Because Breadth First Search explores nodes in increasing order of their distance from the source, the resulting path to any reachable node in the graph will be a shortest path.[4] You can use the same path_to() function to recover the path from src to any node in the graph reachable from src. As visualized in Figure 7-7, Breadth First Search methodically explores the graph.

The queue maintains the search space in order of distance from the source; the nodes are shaded in the queue based on their distance from the source. Whenever a dead end is encountered, no new nodes are enqueued. Note that the target node (2, 2) is added to the queue within the for loop, but the visualization shows the moment when it is dequeued by the outer while loop. All nodes fewer than 2 steps away from the source have been processed, and the very last node in the queue is 3 steps away from the source.

Just for fun, I'll show a third approach for solving rectangular mazes *that takes into account how far away a node is from the target*. Both Depth First Search and Breadth First Search are *blind searches*: that is, they complete their search with only local information about adjacent nodes. The field of artificial intelligence has developed numerous path-finding algorithms that can more effectively complete a search when provided with information about the application domain.

---

4 There may be multiple paths of the same length, but Breadth First Search will discover one of these paths for which none can be shorter.

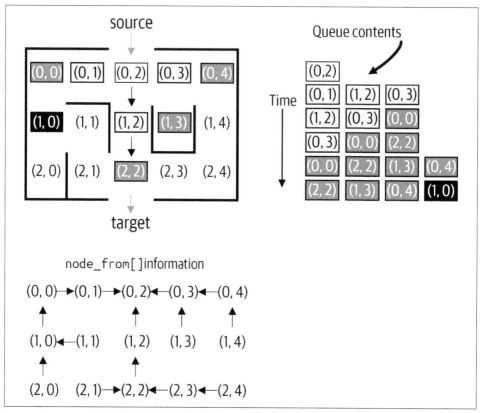

Figure 7-7. Breadth First Search finds shortest path to each node

A Guided Search explores nodes *in order of their shortest physical distance from the target*; to do this, I need to determine how far a node is from the target. Let's define the *Manhattan distance* between two cells in a maze as the sum of the number of rows and columns separating the two nodes.[5] For example, node (2, 0) in the lower-left-hand corner of the sample maze is four steps away from node (0, 2), because it is both two rows away and two columns away.

Given the three adjacent nodes to the source node (0, 2) in the sample maze, Guided Search would first explore (1, 2) because it is only one step away from the target of (2, 2); the other two adjacent nodes are both three steps away using the Manhattan distance. For this idea to work, the nodes being explored need to be stored using a data structure that allows you to retrieve the node closest to the target.

---

5 So called because in a city with a grid street layout, you can't move diagonally, only up, down, left, and right.

One common trick is to use a max priority queue, as presented in Chapter 4, by defining the priority of a node to be *the negative of the Manhattan distance from node to target*. Consider two nodes, where node u is ten steps away from the target, and node v is five steps away. If these nodes are stored in a max priority queue with (u, -10) and (v, -5), then the node with larger priority is v, which is the one closer to the target. The structure of Listing 7-6 is identical to the Breadth First and Depth First Search code, but it uses a priority queue to store the active search space of nodes to explore.

Listing 7-6. A Guided Search using Manhattan distance to control search

```
def guided_search(G, src, target):          ❶
  from ch04.heap import PQ
  marked = {}                                ❷
  node_from = {}                             ❸

  pq = PQ(G.number_of_nodes())               ❹
  marked[src] = True
  pq.enqueue(src, -distance_to(src, target)) ❺

  while not pq.is_empty():                    ❻
    v = pq.dequeue()

    for w in G.neighbors(v):
      if not w in marked:
        node_from[w] = v                      ❼
        marked[w] = True
        pq.enqueue(w, -distance_to(w, target))  ❽

  return node_from                            ❾

def distance_to(from_cell, to_cell):
  return abs(from_cell[0] - to_cell[0]) + abs(from_cell[1] - to_cell[1])
```

❶ Conduct a Guided Search over graph, G, starting from source node, src, knowing the target node to locate.

❷ The marked dictionary records nodes that have already been visited.

❸ Record how search got to each node: node_from[w] is the prior node working backward to src.

❹ Using the heap-based priority queue, you must pre-allocate sufficient space to include, potentially, all nodes in the graph.

❺ Mark and place src node into max priority queue to start the search using as its priority the negative of its distance to target.

❻ If the Guided Search has not yet completed, the node *closest* to `target` is the next node to explore.

❼ For each unmarked node, w, adjacent to v, remember that to get to w, the search came from v.

❽ Place w into its appropriate location in the priority queue, using *the negative of the Manhattan distance* as priority, and mark it so it isn't visited multiple times.

❾ Return the *structure of the search* that records for each node, v, the prior node from a search initiated at `src`.

The intelligence guiding the search is the `distance_to()` function, which computes the Manhattan distance between two nodes.

There is no guarantee that Guided Search will find the shortest path, plus it presumes in advance a single target to guide its search. In particular, it cannot outperform Breadth First Search, which is guaranteed to locate the shortest path not just to the target but to every node in the graph reachable from the source: in doing so, however, Breadth First Search might explore much more of the graph. The hope is that Guided Search reduces unnecessary searches on random maze graphs. Figure 7-8 presents a side-by-side comparison of these three search algorithms on the same maze.

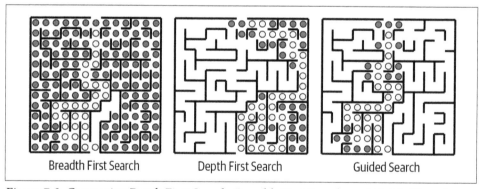

Breadth First Search   Depth First Search   Guided Search

Figure 7-8. Comparing Depth First Search, Breadth First Search, and Guided Search

Breadth First Search will likely explore the most nodes because of its methodical nature. For Depth First Search to discover the shortest path, it must repeatedly choose the right direction to pursue, which is unlikely. There is no guarantee that Guided Search will compute the shortest path between source and target, but as shown in Figure 7-8, it reduces side explorations because it can aim toward the target.

These search algorithms use a `marked` dictionary to ensure each of the N nodes is visited only once. This observation suggests that the runtime performance of each algorithm is $O(N)$, but to confirm this you must validate the performance of the individual operations. With a `Stack`, its operations are performed in constant time—`push()`, `pop()`, and `is_empty()`. The only remaining concern is the efficiency of the `for w in G[v]` loop, which returns the adjacent nodes to v. To classify the performance of this `for` loop, you need to know how the graph stores edges. There are two options, visualized in Figure 7-9, for the maze from Figure 7-3: an adjacency matrix and an adjacency list.

Adjacency matrix

An *adjacency matrix* creates a two-dimensional $N \times N$ matrix, `M`, with N^2 Boolean entries. Each node, u, is assigned an integer index, u_{idx}, from 0 to $N - 1$. If $M[u_{idx}][v_{idx}]$ is `True`, then there is an edge from u to v. These edges are shown as a shaded box, where u is the row label and v is the column label. With an adjacency matrix, retrieving all adjacent nodes for a node, u, requires $O(N)$ runtime performance to check each entry in `M` for u, *regardless of how many adjacent nodes to u actually exist*. Since the `while` loop executes N times, and now the inner `for` loop requires $O(N)$ performance to check N entries in `M`, this means that the search algorithm is classified as $O(N^2)$.

Adjacency list

An *adjacency list* uses a symbol table that associates for each node, u, a *linked list of adjacent nodes*. Retrieving all adjacent nodes for a node, u, requires runtime performance directly proportional to d, where d is the *degree* of node u, or the number of adjacent nodes to u. There is no predetermined ordering for these adjacent nodes, since that is based on how the edges were added to the graph. These linked lists are visualized for each node, u, in Figure 7-9. With an adjacency list, some nodes have few adjacent nodes, while others have a large number of adjacent nodes.

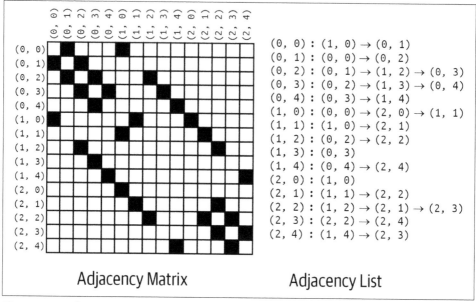

Figure 7-9. Adjacency matrix versus adjacency list representation

The code fragment in Listing 7-7 shows that the runtime performance of Depth First Search is based on E, the number of edges in the graph, *when an adjacency list is used for representation*. Instead of counting the number of *nodes* being processed, count the number of *edges* being processed.

Listing 7-7. Code fragment showing performance based on the number of edges

```
while not stack.is_empty():
  v = stack.pop()
  for w in G[v]:
    if not w in marked:
      marked[w] = True
      stack.push(w)
      ...
```

You have already seen that each node, *v*, can be inserted into the stack only once. This means the if statement *will execute once for every adjacent node to v*. If you consider a graph with just two nodes, *u* and *v*, with a single edge (*u*, *v*), then the if statement will execute *twice*, once when processing the adjacent nodes to *u* and once when processing the adjacent nodes to *v*. So for an undirected graph, the number of times that the if statement executes is 2 × E, where E is the number of edges in the graph.

Put all this together—up to N push() and pop() invocations and 2 × E invocations of the if statement—and you can declare that the runtime performance of these search algorithms when using adjacency lists is O(N + E), where N is the number of nodes and E is the number of edges. The same is true for Breadth First Search, which uses a queue instead of a stack but exhibits the same runtime performance.

In some ways, these results are actually compatible; specifically, in an undirected graph with N nodes, E is less than or equal to N × (N – 1)/2 edges.[6] Regardless of whether the graph is stored using an adjacency matrix or an adjacency list, searching over graphs with a high number of edges will be proportional to N × (N – 1)/2, or O(N^2), so in the *worst case* it will be O(N^2).

Guided Search, however, relies on a priority queue to maintain the nodes closest to the designated target node. The enqueue() and dequeue() operations are O(log N) in the *worst case*. Since these methods are called N times, and each edge is visited twice, the *worst case* performance for Guided Search is O(N log N + E).

Directed Graphs

Graphs can also model problems where the relationship between two nodes is directional, typically represented as an edge with an arrow. A directed edge (u, v) only declares that v is adjacent to u: this edge does not make u adjacent to v. In this edge, u is the *tail*, while v is the *head*—this is easy to remember because the arrow head is adjacent to v. The graph in Figure 7-10 contains the edge (B3, C3) but not (C3, B3).

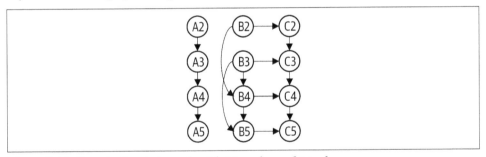

Figure 7-10. Sample directed graph with 12 nodes and 14 edges

The Depth First and Breadth First searches are still relevant for directed graphs: the only difference is that with the edge (u, v), v is adjacent to u, but the opposite is only true if the graph contains a separate edge (v, u). With directed graphs, many algorithms are simplified when using recursion, as shown in Listing 7-8. This code includes many familiar elements from the nonrecursive implementation.

---

6 The triangle numbers make yet another appearance!

Listing 7-8. Recursive implementation of Depth First Search on a directed graph

```
def dfs_search(G, src):        ❶
  marked = {}                  ❷
  node_from = {}               ❸

  def dfs(v):                  ❹
    marked[v] = True           ❺
    for w in G[v]:
      if not w in marked:
        node_from[w] = v       ❻
        dfs(w)                 ❼

  dfs(src)                     ❽
  return node_from             ❾
```

❶ Conduct a Depth First Search over graph G starting from source node, src.

❷ The marked dictionary records nodes that have already been visited.

❸ Record how dfs() found each node: node_from[w] is the prior node working backward to src.

❹ Recursive method to continue search from an unmarked node, v.

❺ Be sure to mark that v has been visited.

❻ For each unmarked node, w, adjacent to v, remember that to get to w, the search came from v.

❼ In the recursive case, continue search in direction of unmarked node, w. When recursive call ends, continue with for loop over w.

❽ Invoke the initial recursive call on source node, src.

❾ Return the *structure of the search* that records for each node, v, the prior node from a search initiated at src.

A recursive algorithm remembers its partial progress using the recursive call stack, so there is no need for a Stack data type.

For every dfs(v) in the recursive call stack (where v is different with each invocation), node v is part of the active search space. In the base case of dfs(v), node v has no unmarked adjacent nodes, so it performs no work. In the recursive case, for each unmarked adjacent node, w, a recursive dfs(w) invocation is launched. When it returns, the `for` loop over w continues, trying to find additional unmarked nodes adjacent to v to explore with `dfs()`.

 As mentioned earlier in this book, Python limits the recursion depth to 1,000, which means some algorithms will not work for large problem instances. For example, a 50 x 50 rectangular maze has 2,500 cells. A Depth First Search will likely exceed the recursion limit. Instead, use a `Stack` data type to store the search progress.[7] The resulting code is often more complicated to understand, so for the rest of this chapter, I use a recursive Depth First Search.

Directed graphs can model application domains that have different problems to solve. Figure 7-11 depicts a small spreadsheet business application containing cells uniquely identified by a column and a row; cell B3 contains the constant 1, and this means the value of B3 is 1. The left image in Figure 7-11 shows the view presented to the user; the middle image shows the actual contents of each cell, including formulas. A cell can contain formulas that refer to other constants, and possibly the values computed from other formulas. These formulas are represented using infix expressions, which you learned in Chapter 6. For example, cell A4 contains the formula "= (A3 + 1)". Now, A3 contains the formula "= (A2 + 1)", which can be computed as 1 (since A2 contains the value 0), which means that A4 is computed to be 2. The code for this sample spreadsheet application is contained in the repository.

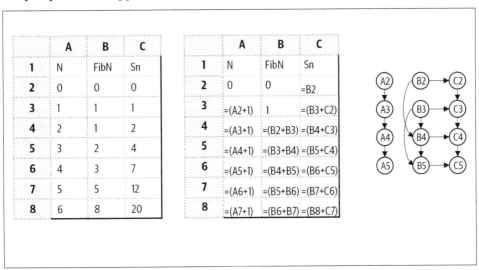

Figure 7-11. Sample spreadsheet with underlying directed graph

---

7 Find the stack-based Depth First Search in `ch07.search` in the code repository.

This spreadsheet computes increasing values for N in column A, while column B contains the first seven Fibonacci numbers.[8] Column C contains the *running total* of the first N Fibonacci numbers (for example, the 12 in cell C7 represents the sum of 0 + 1 + 1 + 2 + 3 + 5). The right image in Figure 7-11 represents the directed graph that captures the relationships between the cells. For example, there is an edge from A2 to A3, reflecting that the value for A3 must change when A2 changes; another way to phrase this relationship is that the value for A2 must be known before the value for A3 can be computed.

In a spreadsheet, if cell C2 contains the formula "= B2" and cell B2 contains the formula "= C2", these cells refer to each other, resulting in a *circular reference*, which is an error. Using the terminology of directed graphs, this situation would be a *cycle*, that is, a sequence of directed edges that starts at a node, n, and returns to n. Every spreadsheet program checks for cycles to make sure that its cells can be properly computed without error. Returning to Figure 7-11, the cells containing constants do not require any computations. The value for A3 needs to be computed before the value for A4 (which is subsequently needed when computing A5). The relationship between the B and C cells is more complicated, making it harder to know the order in which these cells should be computed, let alone whether a cycle even exists.

To operate safely, a spreadsheet application can maintain a directed graph of references between its cells to record dependencies between the cells. Whenever the user changes the contents of a cell, the spreadsheet must remove edges from this graph if the cell had formerly contained a formula. If the changed cell introduces a new formula, then the spreadsheet adds edges to capture the new dependencies in the formula. For example, given the spreadsheet from Figure 7-11, the user could mistakenly create a cycle by changing the contents of cell B2 to be the formula "= C5". With this change, a new edge, C5 → B2, would be added to the directed graph, leading to several cycles; here is one: [B2, B4, B5, C5, B2].

Given a directed graph, Depth First Search can determine whether a cycle exists in the directed graph. The intuition is that when Depth First Search finds a marked node *that is still part of the active search space*, a cycle exists. When dfs(v) returns, you are assured that *all nodes that are reachable from v have been marked*, and v is no longer part of the active search space.

---

8 Recall from Chapter 5 that these are the numbers 0, 1, 1, 2, 3, 5, and 8, computed by adding two successive terms.

A Depth First Search exploring a directed graph with just three nodes can encounter a marked node in a graph without a cycle. Initiating dfs() from node a could explore to b and finally c, which is a dead end. When the search returns to process the remaining adjacent nodes to a, although c is marked, no cycle exists.

This Cycle Detection algorithm differs from the other algorithms in this chapter because there is no initial dedicated source node from which the exploration starts. The question is whether a cycle exists *anywhere* in the graph, so the algorithm, described in Listing 7-9, has to investigate potentially every node in the graph.

Listing 7-9. Detecting cycles in a directed graph using Depth First Search

```
def has_cycle(DG):
  marked = {}
  in_stack = {}

  def dfs(v):                                    ❶
    in_stack[v] = True                           ❷
    marked[v] = True                             ❸

    for w in DG[v]:
      if not w in marked:
        if dfs(w):                               ❹
          return True
      else:
        if w in in_stack and in_stack[w]:        ❺
          return True

    in_stack[v] = False                          ❻
    return False

  for v in DG.nodes():                           ❼
    if not v in marked:
      if dfs(v):                                 ❽
        return True
  return False
```

❶ Conduct a Depth First Search over graph, DG, starting from v.

❷ in_stack records nodes that are in the recursive call stack. Mark that v is now part of a recursive call.

❸ The marked dictionary records nodes that have already been visited.

❹ For each unmarked node, w, adjacent to v, initiate recursive dfs() on w, and if True is returned, a cycle is detected, so it returns True as well.

❺ If a node, w, is marked as visited, it could still be in our call stack—if it is, then a cycle has been detected.

❻ Equally important, when the dfs() recursive call ends, set in_stack[v] to False since v is no longer on the call stack.

❼ Investigate each unmarked node in the directed graph.

❽ If invoking dfs(v) on a node, v, detects a cycle, return True immediately.

As dfs() recursive calls execute, more of the graph is explored until eventually each node is marked—even those with no edges.

If you want to also compute the actual cycle, try the challenge exercise at the end of this chapter that modifies has_cycle() to compute and return the first detected cycle in a directed graph. Figure 7-12 visualizes the recursive execution of dfs(). Each discovered node is eventually marked, but only nodes *in the active search space*—those nodes where in_stack[] is True—are highlighted as the recursion proceeds and unwinds. The image shows the moment in the recursion when the cycle [a, b, d, a] is detected. When exploring the adjacent nodes to d, the marked node, a, is encountered, but this doesn't immediately mean that a cycle exists. The algorithm must check whether in_stack[a] is True to confirm that a cycle exists.

The final recursive invocation of dfs(d) has not yet ended, which is why in_stack[d] is still True. To summarize, when the recursive dfs() function encounters a node, n, that has already been marked *and* in_stack[n] is True, a cycle has been found.

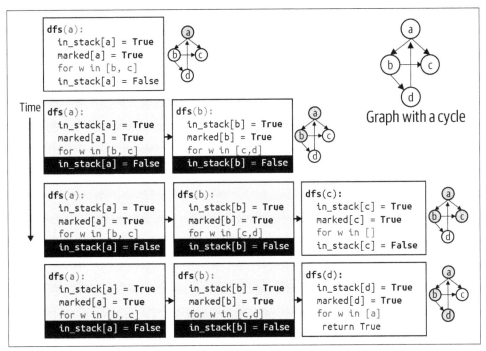

Figure 7-12. Visualizing execution of Depth First Search for Cycle Detection

Assuming that a spreadsheet contains no circular references, in what order should its cells be computed? Returning to the spreadsheet example in Figure 7-11, the cells that contain constants (such as A1) are not involved in any computation, so they do not matter. The cell B4 contains a formula that directly depends on both B2 and B3, so these two cells must be computed before B4. One linear ordering that works is:

B2, C2, B3, C3, B4, C4, B5, C5, A2, A3, A4, A5

The preceding ordering is one possible result from `topological_sort()`, shown in Listing 7-10. The structure of this algorithm is identical to the Cycle Detection algorithm described earlier. It relies on a recursive Depth First Search to explore the graph. When `dfs(v)` is about to return from its recursive invocation, all nodes that are reachable from *v* have been `marked`. This means that `dfs()` has already visited all "downstream" nodes that are dependent on *v*, so it adds *v* to the growing list of nodes (in reverse order) whose dependencies have been processed.

Listing 7-10. Topological sort over the directed graph

```
def topological_sort(DG):
  marked = {}
  postorder = []                          ❶

  def dfs(v):                             ❷
    marked[v] = True                      ❸
    for w in DG[v]:
      if not w in marked:
        dfs(w)                            ❹
    postorder.append(v)                   ❺

  for v in DG.nodes():
    if not v in marked:                   ❻
      dfs(v)

  return reversed(postorder)              ❼
```

❶ Use a list to store (in reverse order) a linear ordering of nodes to be processed.

❷ Conduct a Depth First Search over DG starting from v.

❸ The marked dictionary records nodes that have already been visited.

❹ For each unmarked node, w, adjacent to v, recursively explore dfs(w).

❺ When dfs(v) gets to this key step, it has fully explored all nodes that (recursively) depend on v, so append v to postorder.

❻ Ensure that all unmarked nodes are visited. Note that each time dfs(v) is invoked, a different subset of graph DG is explored.

❼ Because the list holds the linear ordering in reverse order, return its reverse.

This code is nearly identical to the Cycle Detection algorithm, except it maintains the postorder structure instead of in_stack[ ]. Using a similar runtime analysis, you can see that each node has one chance to be explored with dfs(), and the inner if statement is executed once for every directed edge in the graph. Since appending to a list has constant time performance (see Table 6-1), this guarantees that the runtime performance of Topological Sort is $O(N + E)$, where N is the number of nodes and E is the number of edges.

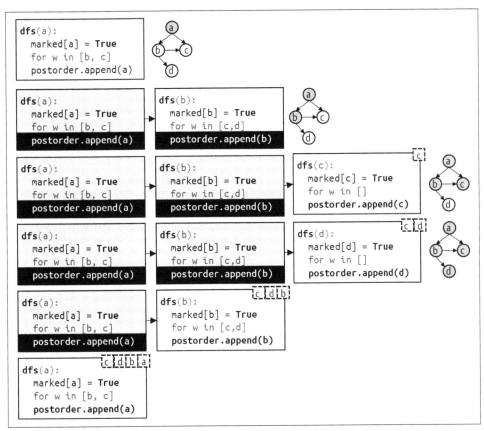

Figure 7-13. Visualizing execution of Depth First Search for Topological Sort

When each `dfs()` completes its execution in Listing 7-10, as shown visually in Figure 7-13, `postorder` contains the ordered list of nodes whose dependencies are satisfied; these are shown in dashed boxes. This list is reversed and returned at the end of `topological_sort()`. When the spreadsheet application loads a spreadsheet document, it can recompute the cells in the order determined by Topological Sort.

Graphs with Edge Weights

Some application domains modeled using graphs have a numeric value associated with each edge, typically called the *weight* of an edge. These edge weights can appear in undirected or directed graphs. For now, assume that all edge weights are positive values greater than 0.

The Stanford Large Network Dataset Collection (*https://oreil.ly/ hXqcg*) contains some large data sets for social networks. Computer scientists have studied the "traveling salesman problem" (TSP) for decades, and numerous data sets are available (TSPLIB (*https:// oreil.ly/MdMWm*)). A large highway data set is available at Travel Mapping Graph Data (*https://oreil.ly/qWYsr*). I'd like to thank James Teresco for graciously providing the Massachusetts highway data set (*https://oreil.ly/wlEy2*).

From the data set of highway fragments in Massachusetts, let's create a graph where each node in the graph represents a *waypoint* from the data set, represented by a (latitude, longitude) pair of values. For example, one waypoint is the intersection of highways I-90 and I-93 in Boston. It is identified by a latitude value of 42.34642 (which means it is north of the equator) and a longitude value of –71.060308 (which means it is west of Greenwich, England). An edge between two nodes represents a highway fragment: the weight of the edge is the length of the highway fragment in miles. Different roads connect these waypoints together, leading to the highway infrastructure shown in Figure 7-14.

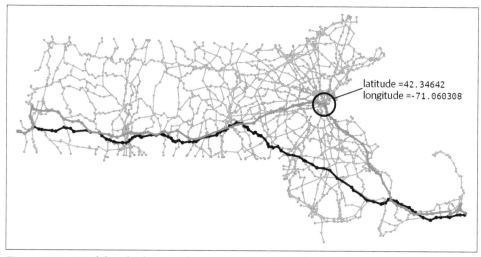

Figure 7-14. Modeling highway infrastructure in Massachusetts

To determine the shortest route (in terms of total mileage) from the westernmost highway in Massachusetts (on the New York Border) to the easternmost highway (on Cape Cod), start with Breadth First Search to compute a 236.5-mile path (highlighted in Figure 7-14). This 99-edge path passes through the identified waypoint on highway I-90/I-93 in Boston and is the shortest path (from the source to the target) in terms of *the total number of edges*. But is it the shortest total path *in accumulated mileage* when considering edge weights? It turns out the answer is no.

We know Depth First Search offers no guarantee on path length: Figure 7-15 contains the wandering 485.2-mile journey with 267 edges produced by Depth First Search. A Guided Search algorithm makes a poor decision early in its journey (not shown here) to compute a 245.2-mile path with 141 edges. The other visualized path in Figure 7-14, which steadily progresses in a southeasterly direction, contains 128 edges and requires only 210.1 miles; Dijkstra's algorithm shows how to compute this solution.

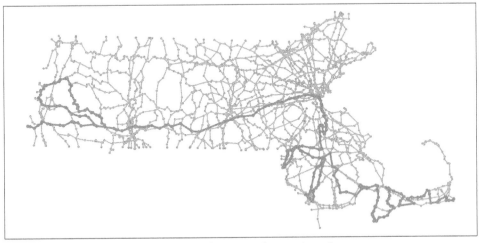

Figure 7-15. Inefficient path resulting from Depth First Search

Dijkstra's Algorithm

Edsger Dijkstra (pronounced DIKE-stra), a scientist from the Netherlands, was one of the intellectual founders of the discipline of computer science, and his algorithms are as elegant as they are insightful. Dijkstra developed an algorithm that computes the shortest path *of accumulated edge weights* from a designated source node to all reachable nodes in a weighted graph. This problem is known as the "single-source shortest path" problem. Given an undirected (or directed) graph, G, with non-negative *weights* associated with each edge,[9] Dijkstra's algorithm computes dist_to[] and edge_to[] structures, where dist_to[v] is the length of the shortest accumulated path from the source node to v, and edge_to[] is used to recover the actual path.

A sample weighted, directed graph is shown in Figure 7-16. The edge (a, b) has a weight of 6. There is an edge from a to c with weight of 10, but a path from a to b (with weight of 6) to c (with weight of 2) has an accumulated total of 8, which

---

9 This allows for some edge weights to be zero. If any edge weight is negative, you will need the Bellman–Ford algorithm (presented later in this chapter).

represents a shorter path. The shortest path from a to c contains two edges and a total weight of 8.

When the graph is directed, it may be impossible to construct a path between two nodes. The shortest distance from b to c is 2, but the shortest distance from b to a is infinity because there is no way to construct a path using the existing edges.

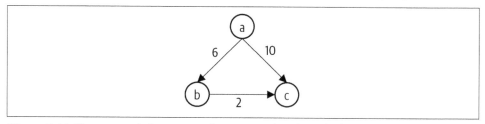

Figure 7-16. Shortest path from a to c has accumulated total of 8

Dijkstra's algorithm requires an abstract data type known as the *indexed min priority queue*, designed specifically to work with graph algorithms. The indexed min priority queue extends the priority queue data type introduced in Chapter 4. An indexed min priority queue associates a priority with each value. The min priority queue is constructed with an initial storage based on N, where N is the number of nodes in the graph being processed. The dequeue() operation removes the value whose priority *is smallest numerically*, in contrast to the max priority queue discussed in Chapter 4.

The most important operation is decrease_priority(value, lower_priority) that efficiently *reduces the priority of* value *to a lower priority*. In effect, decrease_prior ity() can adjust the priority of an existing value so that it *could move ahead of other values in the priority queue*. The priority queue implementations described earlier are unable to provide an efficient decrease_priority() function, since they would have to search the entire priority queue in O(N) to locate the value whose priority changes.

> The indexed min priority queue is typically restricted to use only integer values ranging from 0 to N – 1, which allows it to store and access data easily in arrays. I use a Python dictionary to remove this restriction.

As shown in Listing 7-11, the structure and functionality of IndexedMinPQ is nearly the same as the heap-based max priority queue presented in Chapter 4. Instead of storing Entry objects, IndexedMinPQ stores two lists: values[n] stores the value for the nth item in the heap, while priorities[n] stores its associated priority. The swim() and sink() methods are identical to the heap-based priority implementation (shown in Listing 4-2 and Listing 4-3), so they are omitted. The primary change is a location dictionary that stores the index position in these lists for each value in

IndexedMinPQ. This extra information allows you to determine the location in the heap for any value in amortized constant O(1) time, using the results of hashing (discussed in Chapter 3).

The changes to swap() ensure that whenever two items in the heap are swapped, their respective entries in location are updated. This way IndexedMinPQ can efficiently locate any value in the priority queue.

Listing 7-11. Structure of an indexed min priority queue

```
class IndexedMinPQ:
  def less(self, i, j):                                        ❶
    return self.priorities[i] > self.priorities[j]

  def swap(self, i, j):
    self.values[i],self.values[j] = self.values[j],self.values[i]  ❷
    self.priorities[i],self.priorities[j] = self.priorities[j],self.priorities[i]

    self.location[self.values[i]] = i                          ❸
    self.location[self.values[j]] = j

  def __init__(self, size):
    self.N          = 0
    self.size       = size
    self.values     = [None] * (size+1)                        ❹
    self.priorities = [None] * (size+1)

    self.location   = {}                                       ❺

  def __contains__(self, v):                                   ❻
    return v in self.location

  def enqueue(self, v, p):
    self.N += 1

    self.values[self.N], self.priorities[self.N] = v, p        ❼
    self.location[v] = self.N                                  ❽
    self.swim(self.N)
```

❶   Because this is a min priority queue, item i is of lower priority than item j *if its priority is a larger numeric value.*

❷   swap() switches the values and priorities for items i and j.

❸   swap() updates the respective locations for items i and j.

❹   values stores the value of the nth item; priorities stores the priorities of the nth item.

❺   `location` is a dictionary that returns the index position into `values` and `priori``ties` for each value that is enqueued.

❻   Unlike a traditional priority queue, an indexed min priority queue can inspect `location` to determine in amortized O(1) time whether a value is being stored in the priority queue.

❼   To enqueue a (v, p) entry, place v in `values[N]` and p in `priorities[N]`, which is next available bucket.

❽   `enqueue()` must also associate this new index location with v before invoking `swim()` to guarantee the heap-ordered property.

As you should expect with a heap, `enqueue()` first stores the value, v, and its associated priority, p, at the end of the `values[]` and `priorities[]` lists, respectively. To fulfill its obligation to `IndexedMinPQ`, it also records that value v is stored in index location N (recall that a heap uses 1-based indexing to make the code easier to understand). It invokes `swim()` to ensure the heap-ordered property for `IndexedMinPQ`.

Through its `location[]` array, `IndexedMinPQ` can find the location in the heap for any value that it stores. The `decrease_priority()` method shown in Listing 7-12 can move any value in the `IndexedMinPQ` *closer to the front of the priority queue.* The only restriction is that you can only *decrease* the numeric value of the priority—which makes it potentially more important—and swim the item up into its proper location.

Listing 7-12. Decreasing priority for a value in `IndexedMinPQ`

```
def decrease_priority(self, v, lower_priority):
  idx = self.location[v]                          ❶
  if lower_priority >= self.priorities[idx]:      ❷
    raise RuntimeError('...')

  self.priorities[idx] = lower_priority           ❸
  self.swim(idx)                                  ❹
```

❶   Find the location, `idx`, in the heap where v is found.

❷   If `lower_priority` is actually not lower than the existing priority in `priori``ties[idx]`, raise a `RuntimeError`.

❸   Change the priority for value v to be the lower priority.

❹   Reestablish heap-ordered property if necessary by swimming this value up.

dequeue() removes the value with smallest priority value (which means it is the most important). The IndexedMinPQ implementation is more complicated, because it has to properly maintain the location dictionary, as shown in Listing 7-13.

Listing 7-13. Removing highest-priority value in IndexedMinPQ

```
def dequeue(self):
  min_value = self.values[1]                              ❶

  self.values[1] = self.values[self.N]                    ❷
  self.priorities[1] = self.priorities[self.N]
  self.location[self.values[1]] = 1

  self.values[self.N] = self.priorities[self.N] = None    ❸
  self.location.pop(min_value)                            ❹

  self.N -= 1                                             ❺
  self.sink(1)
  return min_value                                        ❻
```

❶ Remember min_value, the value with highest priority.

❷ Move the item at location N to the top-level location 1 and ensure that location records the new index position for this value.

❸ Remove all trace of the former min_value being removed.

❹ Remove min_value entry from location dictionary.

❺ Reduce number of entries *before* invoking sink(1) to reestablish heap-ordered property.

❻ Return the value associated with entry of highest priority (which is the one with smallest magnitude).

The IndexedMinPQ data structure ensures the invariant that if v is a value stored by the priority queue, then location[v] points to an index location, idx, such that values[idx] is v and priorities[idx] is p, where p is the priority for v.

Dijkstra's algorithm uses an IndexedMinPQ to compute the length of a shortest path from a designated src node to any node in a graph. The algorithm maintains a dictionary, dist_to[v], to record the length of the shortest known computed path from src to each v in the graph: this value may be infinite for nodes not reachable from src. As the algorithm explores the graph, it looks for two nodes, *u* and *v*, connected by an edge with weight of wt such that dist_to[u] + wt < dist_to[v]: in other words, the distance from src to v is shorter if you follow the path from src to u and then cross to v along the edge (u, v).

Dijkstra's algorithm shows how to find these special edges methodically, similar to the way Breadth First Search uses a queue to explore nodes based on their distance from src (in terms of number of edges). dist_to[v] summarizes the results of the active search, and IndexedMinPQ organizes the remaining nodes to be explored by priority, which is defined as the accumulated length of the shortest path for each node from src. When the algorithm starts, dist_to[src] is 0 because that node is the source, and all other distances are infinity. All nodes are then enqueued in the IndexedMinPQ with priority equal to 0 (for the source node, src) or infinity for other nodes.

Dijkstra's algorithm does not need to mark nodes as having been visited, since the min priority queue contains only those active nodes to be explored. One by one, the algorithm removes from the min priority queue the node whose total accumulated distance is the smallest.

Listing 7-14. Dijkstra's algorithm to solve single-source shortest path problem

```
def dijkstra_sp(G, src):
  N = G.number_of_nodes()

  inf = float('inf')                               ❶
  dist_to = {v:inf for v in G.nodes()}
  dist_to[src] = 0

  impq = IndexedMinPQ(N)                            ❷
  impq.enqueue(src, dist_to[src])
  for v in G.nodes():
    if v != src:
      impq.enqueue(v, inf)

  def relax(e):
    n, v, weight = e[0], e[1], e[2][WEIGHT]         ❺
    if dist_to[n] + weight < dist_to[v]:            ❻
      dist_to[v] = dist_to[n] + weight              ❼
      edge_to[v] = e                                ❽
      impq.decrease_priority(v, dist_to[v])         ❾

  edge_to = {}                                      ❸
  while not impq.is_empty():
    n = impq.dequeue()                              ❹
    for e in G.edges(n, data=True):
      relax(e)

  return (dist_to, edge_to)
```

❶ Initialize dist_to dictionary to infinity for all nodes except src, which is 0.

❷ Enqueue all N nodes into impq to prepare for while loop.

❸ edge_to[v] records the edge terminating at v found during the search.

❹ Find node, n, that has shortest computed path from src. Explore its edges (n, v, weight) to see if a new shortest path to v has been found. networkx requires data = True to retrieve edge weights.

❺ Extract n, v, and weight from the edge (n, v).

❻ If distance to n added to edge weight to v is smaller than best path so far to v, then a shorter path has been found.

❼ Update the shortest known distance to v.

❽ Record the edge (n, v) that brought Dijkstra's algorithm to v along the new shortest path.

❾ Most importantly, reduce priority in impq to new shortest distance so while loop will be able to retrieve node with shortest computed path.

Figure 7-17 presents the first three iterations through the while loop in Listing 7-14. IndexedMinPQ stores each node, n, using as priority the smallest computed distance, dist_to[n], which is shown in a small dashed box attached to each node. With each pass through the while loop, a node, n, is removed from impq to check whether its edges lead to a new shortest path from src to some node, v, by traveling from src to n and then from n to v. This process is called *relaxing an edge*. The IndexedMinPQ prioritizes which nodes are explored first—in this way, Dijkstra's algorithm guarantees that *the shortest path for each node dequeued from* impq *is correct*.

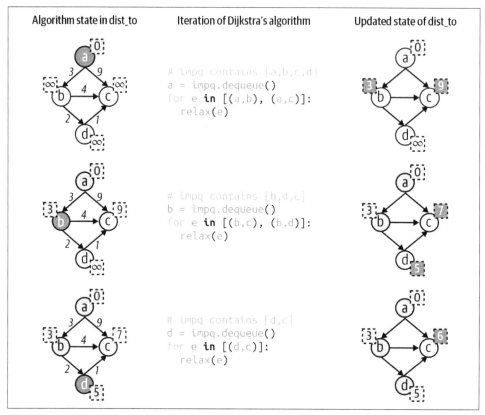

Algorithm state in dist_to	Iteration of Dijkstra's algorithm	Updated state of dist_to

```
# impq contains [a,b,c,d]
a = impq.dequeue()
for e in [(a,b), (a,c)]:
    relax(e)
```

```
# impq contains [b,d,c]
b = impq.dequeue()
for e in [(b,c), (b,d)]:
    relax(e)
```

```
# impq contains [d,c]
d = impq.dequeue()
for e in [(d,c)]:
    relax(e)
```

Figure 7-17. Executing Dijkstra's algorithm on a small graph

The runtime performance of Dijkstra's algorithm is based on several factors:

The cost of enqueuing all N nodes

The first node enqueued is src with priority of 0. All remaining N – 1 nodes are enqueued with infinite priority; since infinity is greater than or equal to every value already in impq, swim() does nothing, resulting in O(N) performance.

The cost of retrieving N nodes from impq

Dijkstra's algorithm dequeues each node from impq. Since impq is stored as a binary heap, dequeue() is O(log N), which means the total time to remove all nodes is O(N log N) in the *worst case*.

The cost of accessing all edges in G

The structure of the graph determines the runtime performance of retrieving all edges in the for e in G.edges() loop. If the graph stores edges using an adjacency matrix, accessing all edges requires O(N^2) performance. If the graph stores edges using an adjacency list, retrieving all edges requires just O(E + N).

The cost of relaxing all E edges

The relax() function is called on all edges in the graph. Each one has a chance to reduce the shortest computed path length to some node, so there can be as many as E invocations of decrease_priority(). This function relies on the swim() binary heap function, whose runtime performance is O(log N). The accumulated time will have a runtime performance of O(E log N).

If the graph stores edges using an adjacency list, Dijkstra's algorithm has an O((E + N) log N) classification. If the graph uses an adjacency matrix instead, the performance is $O(N^2)$. For large graphs the matrix representation is simply inefficient.

Dijkstra's algorithm computes two structures: dist_to[v] contains the length of the shortest path by accumulated edge weights from src to v, while edge_to[v] contains the last edge (*u*, *v*) on the actual shortest path from src to v. The full path from src to each v can be recovered, much like Listing 7-4, except this time following the edge_to[] structure backward, as shown in Listing 7-15.

Listing 7-15. Recovering actual path from edge_to[]

```
def edges_path_to(edge_to, src, target):  ❶
  if not target in edge_to:
    raise ValueError('Unreachable')       ❼

  path = []
  v = target                              ❷
  while v != src:
    path.append(v)                        ❸
    v = edge_to[v][0]                      ❹

  path.append(src)                        ❺
  path.reverse()                          ❻
  return path
```

❶ edge_to[] structure is needed to recover path from src to any target.

❷ To recover the full path, start at target.

❸ As long as v is not src, append v to path, a backward list of nodes found on path from src to target.

❹ Set v to become u, the prior node in the edge (u, v) from edge_to[v].

❺ Once src is encountered, the while loop terminates, so src must be appended to complete the backward path.

❻ Reverse the list so all nodes appear in proper order from src to target.

❼ If target is not in edge_to[], it is not reachable from src.

Dijkstra's algorithm will work *as long as all edge weights are nonnegative*. A graph might have a negative edge, for example, because it represents the refund of a financial transaction. If an edge has a negative edge weight, it could break Dijkstra's algorithm (as shown in Figure 7-18).

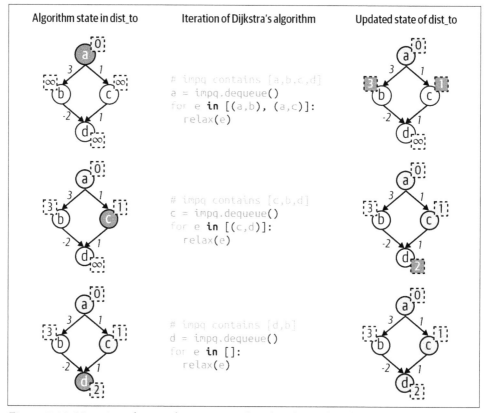

Algorithm state in dist_to	Iteration of Dijkstra's algorithm	Updated state of dist_to
| | ```
# impq contains [a,b,c,d]
a = impq.dequeue()
for e in [(a,b), (a,c)]:
    relax(e)
``` | |
| | ```
# impq contains [c,b,d]
c = impq.dequeue()
for e in [(c,d)]:
    relax(e)
``` | |
| | ```
# impq contains [d,b]
d = impq.dequeue()
for e in []:
    relax(e)
``` | |

Figure 7-18. Negative edge weight in wrong place breaks Dijkstra's algorithm

In Figure 7-18, Dijkstra's algorithm processes three nodes from impq until just b is left. As you can see in the last row, Dijkstra's algorithm has computed its current shortest path from a to d. In its final pass through its while loop (not shown in the figure), Dijkstra's algorithm will remove node b from impq and relax the edge (b, d). Unfortunately, this edge suddenly reveals a shorter path to d. However, Dijkstra's algorithm has already removed node d from impq, finalizing its shortest path computation. Dijkstra's algorithm cannot "go back" and adjust the shortest path, so it fails.

Dijkstra's algorithm can fail with negative edge weights since it assumes that extending an existing path with a new edge will only maintain or increase the total distance
```

from the source. The Bellman–Ford algorithm computes the shortest total distance from src to any other node in the graph, even with negative edge weights, with one exception: if a *negative cycle* exists in the graph, then the concept of a shortest path does not apply. The graph on the left side of Figure 7-19 has two negative edges, but no negative cycle. Using the edge (a, b), the shortest distance from a to b is 1. If you travel over the longer path a → b → d → c → b, the total accumulated edge weight distance is 2, so the shortest path between a and b remains 1. In the graph on the right, however, there is a *negative cycle* between nodes b, d, and c; that is, if you travel the edges clockwise in order from b → d → c → b, the total accumulated edge weight is –2. In this graph, the shortest distance between a and b has no meaning: you can construct a path to make this distance any odd negative number by cycling through the b → d → c → b loop a number of times. The accumulated edge weights for a → b → d → c → b → d → c → b is –3, for example.

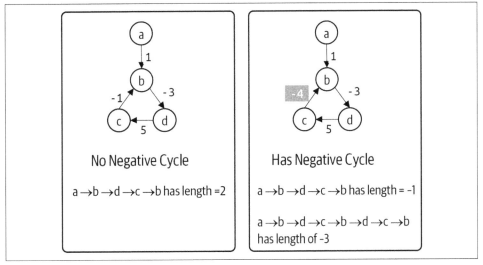

Figure 7-19. Two graphs with negative edge weights, but only one has negative cycle

The Bellman–Ford implementation provides a completely different approach to solving the same single-source shortest path problem. It works even when there are edges with a negative weight. Listing 7-16 shows the Bellman–Ford implementation, which contains many familiar elements from Dijkstra's algorithm. The good news is that you do not have to search for a negative cycle in the graph as you had to do with Topological Sort shown earlier in the chapter. As Bellman–Ford executes, it can detect when a negative cycle exists and raises a runtime exception in response.

Listing 7-16. Bellman–Ford algorithm implementation

```python
def bellman_ford(G, src):
    inf = float('inf')
    dist_to = {v:inf for v in G.nodes()}          ❶
    dist_to[src] = 0
    edge_to = {}                                   ❷

    def relax(e):
        u, v, weight = e[0], e[1], e[2][WEIGHT]
        if dist_to[u] + weight < dist_to[v]:       ❺
            dist_to[v] = dist_to[u] + weight       ❻
            edge_to[v] = e                         ❼
            return True                            ❽
        return False

    for i in range(G.number_of_nodes()):           ❸
        for e in G.edges(data=True):               ❹
            if relax(e):
                if i == G.number_of_nodes()-1:      ❾
                    raise RuntimeError('Negative Cycle exists in graph.')

    return (dist_to, edge_to)
```

❶ Initialize `dist_to` dictionary to infinity for all nodes except `src`, which is 0.

❷ `edge_to[v]` records the edge terminating at v found during the search.

❸ Make N passes over the graph.

❹ For every edge e = (u,v) in the graph, use the same `relax()` concept as Dijkstra's algorithm; see if e improves on an existing shortest path from `src` to v by going through u.

❺ If distance to u added to edge `weight` to v is smaller than best path so far to v, then a shorter path has been found.

❻ Update the shortest known distance to v.

❼ Record the edge (u, v) that brought algorithm to v along the new shortest path.

❽ If `relax()` returns `True`, then a new shortest path was found to v.

❾ Bellman–Ford makes N passes over all E edges. If in the final pass, an edge, e, is found that still reduces the shortest distance from `src` to some v, there must be a negative cycle in the graph.

Why does this algorithm work? Observe that in a graph with N nodes, the longest possible path that can exist in the graph has no more than N − 1 edges. After the for loop over i has made N − 1 iterations attempting to relax any edge in the graph, it must have processed this potentially longest path in the graph: there should no longer be any edge left that relaxes the shortest total distance. For this reason, the for loop iterates N times. If on the final pass, the algorithm is able to relax an edge, then the graph must contain a negative cycle.

All-Pairs Shortest Path

The search algorithms presented in this chapter *start their search from a designated source node.* As its name suggests, the all-pairs shortest path problem asks to compute the shortest possible path of accumulated edge weights between any two nodes, *u* and *v*, in the graph. In an undirected graph, the shortest path from *u* to *v* is the same as the shortest path from *v* to *u*. In both undirected and directed graphs, node *v* may not be reachable from *u*, in which case the shortest path distance would be infinity. In a directed graph, *u* may be reachable from *v* even if *v* is not be reachable from *u*.

The search must return information to be able to recover the actual shortest paths between any *u* and *v*, but this seems incredibly challenging. For the small directed graph in Figure 7-20, it takes time to determine the shortest path between d and c, let alone all possible pairs of nodes. While there is an edge from d to c with weight 7, there is a path d → b → a → c whose total accumulated distance is 6, which is shorter.

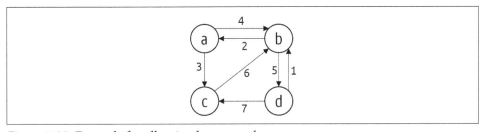

Figure 7-20. Example for all-pairs shortest path

Before delving into the details of an algorithm that solves this problem, consider what the algorithm needs to return as its result. It is similar to what you have seen for Dijkstra's algorithm and the earlier search algorithms:

- dist_to[u][v]—a two-dimensional structure that holds the value of the shortest path between every pair of nodes, *u* and *v*. If there is no path from *u* to *v*, dist_to[u][v] = infinity.

- `node_from[u][v]`—a two-dimensional structure that contains information to make it possible to compute *for any two nodes, u and v,* the actual shortest path between them.

The following insight helps lead to a solution.

Start by initializing `dist_to[u][v]` to be the weight associated with each edge u to v; if no edge exists in the graph, then set `dist_to[u][v]` to be infinity. Also initialize `node_from[u][v]` to be u to record that the last node on the shortest path from u to v is u. Figure 7-21 presents the `node_from[][]` and `dist_to[][]` after initializing these values using the graph from Figure 7-16.

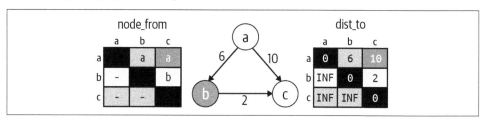

Figure 7-21. Intuition behind the all-pairs shortest path problem

Now imagine that you select k = b and check if you can find any two nodes, u and v, where the path from u to k and then from k to v is shorter than the best known shortest computed distance of `dist[u][v]`. In this small example, `dist[a][b]` is 6, and `dist[b][c]` is 2, which means that the shortest distance from a to c now goes through b and is a total of 8. In addition, you can set `node_from[a][c]` to b to record this fact. This situation is similar to the relaxation computation that was central to Dijkstra's algorithm.

I now have a clear explanation for `node_from[u][v]`: it stores *the last node on the shortest path from node u to v*. It is similar in spirit to the `node_from[]` structures computed by earlier search algorithms, although it is more complicated.

One by one, you set k to each of the nodes in the graph and try to find a pair of nodes, u and v, that could reduce its shortest path distance using the logic in the previous paragraph. Once you know that `dist_to[u][k]` + `dist_to[k][v]` is shorter than `dist_to[u][v]`, you can update the value for `dist_to[u][v]`. You can also set `node_from[u][v]` to equal `node_from[k][v]`.

In other words, since `dist_to[k][v]` has already been computed, you know that `node_from[k][v]` is the last node on the shortest path from k to v—and since the path from u to v now *goes through* k, set `node_from[u][v]` equal to `node_from[k][v]`. To reconstruct the full path, work backward from v to k, then `node_from[u][k]` contains the former node on the shortest path all the way back to u.

These concepts are challenging because they are abstract—I do not compute and store each shortest path between every u and v; rather, I store partial details about the paths to compute them later. Given the graph in Figure 7-20, Figure 7-22 contains the result that an algorithm must compute. The dist_to[][] structure contains the computed shortest distances between any two nodes. The dist_to[a] row, for example, contains the computed shortest distances from a to all other nodes in the graph. In particular, dist_to[a][c] is 3 because the shortest path from a to c is along the (a, c) edge with a total distance of 3. The shortest path from a to d, dist_to[a][d], is along the path a → b → d, whose accumulated total distance is 9.

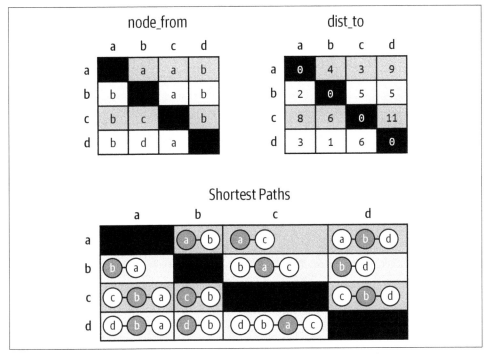

Figure 7-22. *dist_to, node_from, and actual shortest paths for graph in Figure 7-20*

A solution to the all-pairs shortest path problem needs to compute node_from[][] and dist_to[][]. To explain the values in node_from[][], Figure 7-22 shows the resulting shortest paths between each pair of nodes, u and v. In each of the shortest paths between u and v, the *second-to-last node in the shortest path* is highlighted. You can see that this highlighted node corresponds directly to node_from[u][v].

Consider the shortest path from d to c, which is the path d → b → a → c. Decompose this path into a path from d to a followed by the final edge from a to c, and you can see the *recursive solution hidden in these two-dimensional structures*. node_from[d][c] equals a, which means the last node in the shortest path from d to c is the node a.

Next, the shortest path from d to a goes through b, which is why node_from[d][a] equals b.

Floyd–Warshall Algorithm

Now that you are familiar with the all-pairs shortest path problem, I can present the Floyd–Warshall algorithm. The key to the algorithm is its ability to find three nodes, u, v, and k, such that there is a shorter path from u to v by going through k.

In Listing 7-17, Floyd–Warshall initializes node_from[][] and dist_to[][] using just the edge information provided in the initial graph. These initial values are depicted in Figure 7-23.

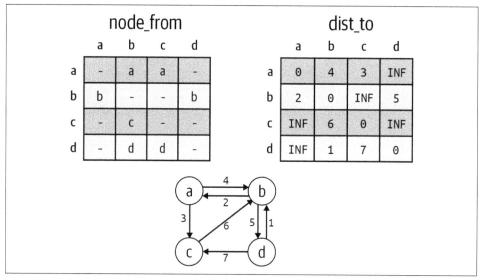

Figure 7-23. Initializing dist_to[][] and node_from[][] based on G

The entry node_from[u][v] is either None (shown as a dash) or u, to reflect the row of the entry. dist_to[u][v] is 0 whenever u is equal to v. When u and v are distinct, dist_to[u][v] is either the weight of the edge from u to v, or infinity (shown by INF) when there is no edge between u and v.

Listing 7-17. Floyd–Warshall Algorithm

```
def floyd_warshall(G):
  inf = float('inf')
  dist_to   = {}                                              ❶
  node_from = {}
  for u in G.nodes():
    dist_to[u]   = {v:inf for v in G.nodes()}                 ❷
    node_from[u] = {v:None for v in G.nodes()}                ❸

    dist_to[u][u] = 0                                         ❹
    for e in G.edges(u, data=True):                           ❺
      v = e[1]
      dist_to[u][v] = e[2][WEIGHT]
      node_from[u][v] = u                                     ❻

  for k in G.nodes():
    for u in G.nodes():
      for v in G.nodes():
        new_len = dist_to[u][k] + dist_to[k][v]               ❼
        if new_len < dist_to[u][v]:
          dist_to[u][v] = new_len                             ❽
          node_from[u][v] = node_from[k][v]

  return (dist_to, node_from)                                 ❾
```

❶ dist_to and node_from will be two-dimensional structures. Each is a dictionary containing sub-dictionaries dist_to[u] and node_from[u].

❷ For each u row, initialize dist_to[u][v] to be infinity to reflect that each node, v, is initially unreachable.

❸ For each u row, initialize node_from[u][v] to be None to reflect that there may not even be a path from u to v.

❹ Make sure to set dist[u][u] = 0 to reflect that the distance from u to itself is 0.

❺ For each edge e = (u,v) from u, set dist_to[u][v] = weight of e to reflect that the shortest distance from u to v is the edge weight for e.

❻ Record that u is the last node on the shortest path from u to v. In fact, it is the only node on the path, which contains just the edge (u, v).

❼ Select three nodes—k, u, and v—and compute new_len, the total path length from u to k added to the path length from k to v.

❽ If `new_len` is smaller than the computed length of the shortest path from u to v, set `dist_to[u][v]` to `new_len` to record the shortest distance, and record that *the last node on the shortest path from u to v is the last node on the shortest path from k to v*.

❾ Return computed `dist_to[][]` and `node_from[][]` structures so actual shortest paths can be computed for any two nodes.

Once the algorithm initializes `node_from[][]` and `dist_to[][]`, the code is quite brief. The outermost `for` loop over k seeks to find two nodes, *u* and *v*, such that the shortest path from *u* to *v* can be shortened by first traveling from *u* to k and then from k to *v*. As k explores more nodes, it eventually tries every possible improvement and ultimately computes the final correct result.

When k is a, its inner `for` loops over u and v discover that for u = b and v = c, a shorter path exists from b to c if it goes through node a. Observe that `dist_to[b][a]` = 2 and `dist_to[a][c]` = 3, whose total of 5 is smaller than infinity (the current computed `dist_to[b][c]` value from Figure 7-23). Not only is `dist_to[b][c]` set to 5, but also `node_from[b][c]` is set to a to reflect the change in the newly discovered shortest path, b → a → c: note that the second-to-last node on the shortest path from b to c is a.

Another way to explain why the algorithm works is to evaluate the values contained in `dist[u][v]`. Before the first pass through the `for` loop over k, `dist[u][v]` records the length of the shortest path from any u to any other v in the graph *that doesn't involve any node other than u and v*. After the first k loop iteration has ended for k = a, `dist[u][v]` records the length of the shortest path from any u to any other v *that can also involve a*. The shortest path from b to c is b → a → c.

When k is b during its second pass through the `for` loop, Floyd–Warshall finds five different pairs of nodes u and v for which the path from u to v is shorter if the path goes through b. For example, the shortest path from d to c had been 7 because of edge (d, c) in the graph. Now, however, the algorithm finds a shorter path that involves b, specifically traveling from d to b (with distance 1) and then traveling from b to c (with distance 5) for a shorter accumulated path of length 6. The final values in `dist[][]` and `node_from[][]` are shown in Figure 7-24.

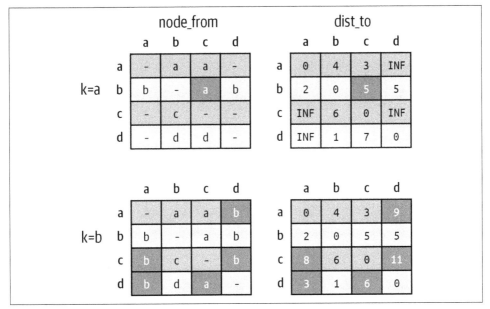

Figure 7-24. Changes to node_from[][] *and* dist_to[][] *after k processes a and b*

You might be surprised that there is no check in Floyd–Warshall to make sure that k, u, and v are distinct nodes. You don't need one, because dist_to[u][u] is initialized to 0. In addition, it would only complicate the code and add unnecessary logic checks.

This algorithm surprisingly uses no advanced data structures but methodically checks all N^3 total (k, u, v) nodes:

- When Floyd–Warshall initializes node_from[][] and dist_to[][], it computes all shortest paths between any *u* and *v* that involve just a single edge.

- After the first pass through k, the algorithm has computed all shortest paths between any *u* and *v* that involve up to two edges, limited to *u*, *v*, and node a.

- After the second pass through k, it has computed all shortest paths between any *u* and *v* that involve *up to three edges*, limited to *u*, *v*, and nodes a *and* b.

Once the outer `for` loop over k has completed processing N nodes, Floyd–Warshall has computed the shortest paths between any *u* and *v* that involve *up to N + 1 edges* and involve any node in the graph. Now, since a path over N nodes can only have N – 1 edges, this means that Floyd–Warshall correctly computes the distance for the shortest paths over all *u* and *v* in the graph.

The code to recover the actual shortest path is shown in Listing 7-18. This code is nearly identical to Listing 7-4, except now it processes a two-dimensional `node_from[][]` structure.

Listing 7-18. Code to recover the shortest path as computed by Floyd–Warshall

```
def all_pairs_path_to(node_from, src, target):     ❶
  if node_from[src][target] is None:
    raise ValueError('Unreachable')                ❼

  path = []
  v = target                                       ❷
  while v != src:
    path.append(v)                                 ❸
    v = node_from[src][v]                           ❹

  path.append(src)                                  ❺
  path.reverse()                                    ❻
  return path
```

❶ `node_from[][]` structure is needed to recover path from `src` to any `target`.

❷ To recover the full path, start at `target`.

❸ As long as v is not `src`, append v to `path`, a backward list of nodes found on path from `src` to `target`.

❹ Set v to become the prior node in the search as recorded by `node_from[src][v]`.

❺ Once `src` is encountered, the `while` loop terminates, so `src` must be appended to complete the backward `path`.

❻ Reverse the list so all nodes appear in proper order from `src` to `target`.

❼ If `node_from[target]` is `None`, `target` is not reachable from `src`.

Summary

Graphs can model a variety of application domains, ranging from geographic data to bioinformatic information to social networks. The edges of a graph can be directed or undirected, and the edges may store numeric weights. Given a graph, there are many interesting questions that naturally arise:

- Is the graph connected? Apply Depth First Search and see if every node in the graph was visited.

- Does a directed graph contain a cycle? Apply Depth First Search and maintain extra state while searching to detect if a cycle exists.

- Given two nodes, u and v, in a graph, what is the shortest path from u to v in terms of the number of edges involved? Apply Breadth First Search to compute a solution.

- Given a weighted graph and starting node, s, what is the shortest path from s to every other node, v, in the graph in terms of accumulated weights of the edges? Apply Dijkstra's algorithm to compute these distances and an edge_to[] structure that can be used to recover the actual paths from s to any reachable node, v.

- If a graph contains negative edge weights—but has no negative cycles—is it still possible to determine the shortest path between a starting node, s, and every other node, u, in terms of accumulated weights of the edges? Apply Bellman–Ford.

- Given a weighted graph, what is the shortest path between any two nodes, u and v, in terms of accumulated weights of the edges? Apply Floyd–Warshall to compute both the distances and a node_from[][] structure that can be used to recover the actual paths.

When working with graphs, do not implement your own data structures to represent these graphs: better to use an existing third-party library, such as NetworkX, so you can benefit immediately from the many algorithms that it provides.

Challenge Exercises

1. Depth First Search can be coded recursively. However, doing so has a weakness when searching large graphs, because Python imposes a recursion limit of around 1,000. Still, for small mazes, you can modify the search to use a recursive search. Modify the skeleton code in Listing 7-19 to recursively invoke Depth First Search. Instead of using a stack to store marked nodes to be removed and processed, only invoke dfs() on marked nodes, and let the recursion unwind to find paths not chosen.

Listing 7-19. Complete recursive implementation for Depth First Search

```
def dfs_search_recursive(G, src):
  marked = {}
  node_from = {}

  def dfs(v):
    """Fill in this recursive function."""

  dfs(src)
  return node_from
```

2. The `path_to()` function to compute the path for Breadth First Search and Depth First Search can be implemented recursively. Implement `path_to_recur sive(node_from, src, target)` as a Python generator that yields the nodes in order from `src` to `target`.

3. Design a `recover_cycle(G)` function that detects when a cycle exists *and returns the cycle*.

4. Design a `recover_negative_cycle(G)` function to augment Bellman–Ford by creating a custom `NegativeCycleError` class extending `RuntimeError` that stores the negative cycle that was discovered in the graph. Start with the offending edge that was relaxed, and try to find a cycle including this edge.

5. Construct a sample directed, weighted graph with N = 5 nodes that requires 4 iterations by Bellman–Ford to properly compute the shortest path from a designated source node. For simplicity, assign each edge a weight of 1. As a hint, it depends on the way that the edges are added to the graph. Specifically, Bellman–Ford processes all edges in order based on how `G.edges()` returns the edges.

6. For randomly constructed N × N mazes, compute the efficiency of Depth First, Breadth First, and Guided Search in reaching the designated target. Do this by revising each search algorithm to (a) stop when it reaches the target, and (b) report the total number of nodes in the `marked` dictionary.

For N equal to powers of 2 ranging from 4 to 128, generate 512 random graphs and compute the average number of marked nodes for each search technique. You should be able to demonstrate that Guided Search is the most efficient, while Breadth First Search is the least efficient.

Now construct a worst case problem instance for Guided Search that forces it to work almost as hard as Breadth First Search. The sample 15 × 15 maze in Figure 7-25 contains walls that form a "U" shape that blocks the path to the exit. Guided Search will have to explore this entire inner space of $(N-2)^2$ cells before "spilling over" one of the edges to find the roundabout path to the exit. The `initialize()` method in `Maze` will be helpful; you will have to manually remove south and east walls to create this shape.

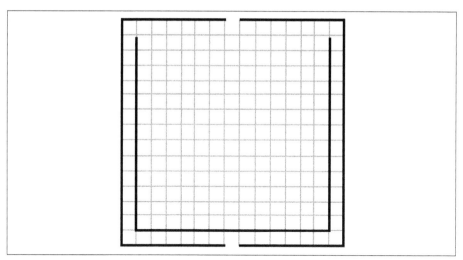

Figure 7-25. Worst case maze for Guided Search

7. A directed graph, DG, with no cycles is called a *directed acyclic graph*, or DAG for short. Dijkstra's algorithm in the *worst case* is classified as $O((E+N) \log N)$, but for a DAG you can compute the single-source, shortest path in $O(E+N)$. First, apply Topological Sort to produce a linear order of the nodes. Second, process each node, n, in linear order, relaxing the edges that emanate from n. There is no need to use a priority queue. Confirm runtime behavior on random *mesh* graphs where each edge has a weight of 1. In the mesh graph in Figure 7-26, the shortest distance from node 1 to node 16 is 6.

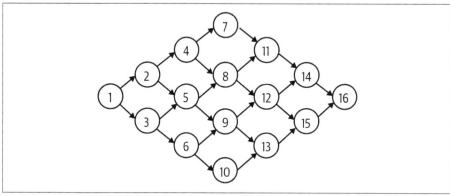

Figure 7-26. A directed, acyclic graph for single-source, shortest path optimization

8. Some drivers prefer to avoid toll roads, such as I-90 in Massachusetts. Given the graph constructed for Massachusetts highways, an edge (u, v) is part of I-90 if the label for both u and v contains 'I-90'. Of the original 2,826 edges, 51 edges are part of I-90: remove these edges from the graph and compute

the shortest distance from the westernmost point in Massachusetts to downtown Boston, whose label (as circled in Figure 7-14) is the string `I-90@134&I-93@20&MA3@20(93)&US1@I-93(20)` representing where six highways converge. With no restrictions, the trip requires 72 edges and a total distance of 136.2 miles. However, if you choose to avoid I-90, the trip requires 104 edges and a total distance of 139.5 miles. Write code to produce these results and output an image file showing the altered route.

Wrapping It Up

My goal in this book was to introduce you to the fundamental algorithms and the essential data types used in computer science. You need to know how to efficiently implement the following data types to maximize the performance of your code:

Bag

A linked list ensures O(1) performance when adding a value. If you choose to use an array, then you need to employ geometric resizing to extend the size of the array so you can guarantee amortized O(1) performance over its average use (though you will still incur an O(N) runtime performance on the infrequent resize events). Note that a bag does not typically allow values to be removed, nor does it prevent duplicate values from being added.

Stack

A linked list can store the values in a stack so push() and pop() have O(1) runtime performance. The stack records the top of the stack to push and pop values.

Queue

A linked list can efficiently store a queue so enqueue() and dequeue() have O(1) runtime performance. The queue records the first and the last node in the linked list to efficiently add and remove values from the queue.

Symbol table

The open addressing approach for symbol tables is surprisingly efficient, with suitable hash functions to distribute the (key, value) pairs. You still need geometric resizing to double the size of the storage, thus effectively making these resize events less frequent.

Priority queue

The heap data structure can store (value, priority) pairs to support `enqueue()` and `dequeue()` in O(log N) runtime performance. In most cases, the maximum number of values to store, N, is known in advance; if not, however, the geometric resizing strategy can be used to minimize the number of resize events.

Indexed min priority queue

Implementing this data type combines the heap data structure with a separate symbol table that stores the index location of each value in the heap. For graph algorithms, it is common to store only integer node labels in the range from 0 to N − 1, where N is the maximum number of values to be stored in the indexed min priority queue. In this case, the separate symbol table can be implemented as an array for extremely efficient look-up. It supports `enqueue()`, `dequeue()`, and `decrease_priority()` in O(log N) runtime performance.

Graph

The adjacency matrix structure is appropriate when all possible edges in the graph exist, which is a common use case for algorithms that compute shortest distances. You can use a two-dimensional array to store an adjacency matrix if the nodes are represented using integers in the range from 0 to N − 1. In most cases, however, an adjacency list structure is more suitable for storing the graphs, and a symbol table is used to associate a bag of adjacent nodes for each node (or, alternatively, a bag of adjacent edges). Any home-built implementation of a graph is going to be insufficient in the long run. Instead, use NetworkX (or any other existing third-party library) to represent graphs efficiently.

The book preface contains a picture that summarizes each of these data types. Throughout the book I have demonstrated how different data structures can efficiently implement the data types, leading to the performance classifications in Table 8-1.

Table 8-1. Performance of data types

Data type	Operation	Classification	Discussion
Bag	size()	O(1)	Use a linked list to store the values for a bag, since prepending a value to the beginning of the linked list is a constant time operation.
	add()	O(1)	
	iterator()	O(N)	
Stack	push()	O(1)	Use a linked list for a stack, pushing new values to the front of the linked list and popping values off the front. If an array is used for storage, constant time performance is possible, but the array might become full.
	pop()	O(1)	
	is_empty()	O(1)	
Queue	enqueue()	O(1)	Use a linked list for a queue, storing references to the first and last node. dequeue the first value by advancing first, while enqueue appends new values after last. If an array is used for storage, constant time performance is possible only with the *circular buffer* technique presented in Chapter 4, but it still can become full.
	dequeue()	O(1)	
	is_empty()	O(1)	
Symbol table	put()	O(1)	Use an array of M linked lists to store N (key, value) pairs in amortized constant time. As more pairs are added, use geometric resizing to double the size of M to increase efficiency. With *open addressing*, a single contiguous array stores all pairs, using linear probing to resolve conflicts. Iterators can return all keys or values. If you also need to retrieve the keys in sorted order, then use a binary search tree where each node stores a (key, value) pair but then the performance of put() and get() become O(log N).
	get()	O(1)	
	iterator()	O(N)	
	is_empty()	O(1)	
Priority queue	add()	O(log N)	A heap data structure can store the (value, priority) pairs, using geometric resizing if storage becomes full. The swim() and sink() techniques from Chapter 4 ensure the performance is O(log N).
	remove_max()	O(log N)	
	is_empty()	O(1)	
Indexed min priority queue	add()	O(log N)	Starting from a heap data structure, store an additional symbol table to look up the location in the heap of any value in O(1) time. Using the symbol table, these operations have O(log N) performance.
	remove_min()	O(log N)	
	decrease_priority()	O(log N)	
	is_empty()	O(1)	

Python Built-in Data Types

The Python implementations are highly tuned and optimized after decades of use. It is impressive how the designers of the Python interpreter constantly seek new ways to gain even a small improvement in performance with each updated release. The *Design and History FAQ* for Python is worth a read.

The four built-in container types in Python are tuple, list, dict, and set:

tuple *data type*

A tuple is an immutable sequence of values that can be processed like a list, except that none of its values can be altered. Use a tuple to return multiple values from a function.

list *data type*

The built-in list data type is the dominant data structure in Python. It is extremely versatile, especially because of Python's slicing syntax that allows programmers to effortlessly work with iterables and select subranges of a list for processing. The list is a general purpose structure, implemented as a variable-length array that uses a contiguous array of references to other values.

dict *data type*

The built-in dict data type represents a symbol table that maps keys to values. All of the concepts in Chapter 3 continue to apply. The Python implementation uses *open addressing* to resolve collisions between keys and also uses storage arrays where M is a power of 2, which is different from how most hashtables are constructed. Each dict internal storage has storage for at least M = 8. This is done so it can store five entries without causing a resize (whereas smaller values of M would require a resize too quickly). It is also optimized to deal with sparse hashtables where most of the storage array is empty. The dict automatically resizes based on a *load factor* of ⅔.

Because Python is open source, you can always inspect the internal implementation.[1] Instead of using linear probing, as described in Chapter 3, a collision at hash code hc will subsequently select the next index to be $((5 \times hc) + 1) \% 2^k$, where 2^k is M, or the size of the storage array. The implementation adds another layer of key distribution using a perturb value that is added to hc. It offers a case study showing how minor mathematical improvements to hash code computations can improve the efficiency of the implementation. Because Python uses a different probing technique, its hashtables can have sizes that are powers of 2, which can eliminate the use of modulo when computing hash codes. This can speed up processing dramatically because the Python interpreter uses a technique known as *bit masking* to perform computations modulo a power of 2.

Specifically, using the bitwise *and* (&) operator, $N \% 2^k$ is equal to $N \& (2^k - 1)$. Table 8-2 shows the difference when timing 10,000,000 computations. The improvement is more pronounced within the Python interpreter (typically written in C), where bitwise *and* is more than five times faster than using modulo.

---

1 For example, you can find the implementation of dict in *https://oreil.ly/jpI8F*.

Table 8-2. Modulo power of 2 is faster with bitwise and, where M is 2^k and M_less is M – 1

Language	Computation	Performance
Python	1989879384 % M	0.6789181 secs
Python	1989879384 & M_less	0.3776672 secs
C	1989879384 % M	0.1523320 secs
C	1989879384 & M_less	0.0279260 secs

set *data type*

A `set` collection contains distinct hashable objects.[2] It is quite common to use a symbol table to act as a set, by simply mapping each key to some arbitrary value, such as 1 or `True`. The internal Python implementation is heavily based on the `dict` approach, but it is worth mentioning that the use cases for a `set` are quite different from a `dict`. Specifically, a `set` is often used for membership testing, to check to see whether a value is contained by the set—and the code is optimized to handle both hits (i.e., the value is contained in the set) and misses (i.e., the value is not contained in the set). In contrast, when working with a `dict`, it is much more common that the key exists within the `dict`.

Implementing Stack in Python

A Python `list` can represent a stack, offering an `append()` function to *push* new values to the end of the list (recall from Table 6-1 that appending to the end of a list is efficient). The `list` type actually has a `pop()` method that removes and returns the last value in the `list`.

Python provides a `queue` module that implements *multi-producer, multi-consumer* stacks that support both "Last-in, first-out" (LIFO) behavior (as regular stacks should) and "First-in, first-out" (FIFO) behavior (as you would expect with a queue). The `queue.LifoQueue(maxSize = 0)` constructor will return a stack that can store a maximum number of values. This stack is meant to be used concurrently, which means that attempting to push a value to a full stack will *block the execution until a value has been popped*. Values are pushed with `put(value)` and popped with `get()`.

---

2 Find its source at *https://oreil.ly/FWttm*.

Invoking the following sequence of commands will freeze the Python interpreter, and you will have to forcefully stop it:

```
import queue
q = queue.LifoQueue(3)
q.put(9)
q.put(7)
q.put(4)
q.put(3)
... blocks until terminated
```

As it turns out, the fastest implementation is deque (pronounced DECK) from the collections module, which stands for "double-ended queue." This data type allows values to be added to (or removed from) either end. The performance of LifoQueue is about 30 times slower than deque, though both provide O(1) runtime performance.

Implementing Queues in Python

A Python list can also represent a queue, offering an append() function to *enqueue* new values to the end of the list. You will need to remove the first element from the list using pop(0) to request removing the element at index position 0 in the list. Using the list in this fashion will become seriously inefficient, as shown in Table 6-1: you must avoid this at all costs.

The queue.Queue(maxSize = 0) function constructs a queue that allows maxSize values to be enqueued, but this should not be your default queue implementation. This queue provides *multi-producer, multi-consumer* behavior, which means that attempting to enqueue a value to a full queue will *block the execution until a value has been dequeued*. Values are enqueued with put(value) and dequeued with get(). Invoking the following sequence of commands will freeze the Python interpreter, and you will have to forcefully stop it:

```
>>> import queue
>>> q = queue.Queue(2)
>>> q.put(2)
>>> q.put(5)
>>> q.put(8)
... blocks until terminated
```

If you need a simple queue implementation, you could use the queue.SimpleQueue() function,[3] which provides the behavior for a queue with a simplified interface. Use put(value) to enqueue value to the end of the queue, and get() to retrieve the value at the head of the queue. This queue is much more powerful than most programmers need, because it handles not only thread-safe concurrent code, but also more compli-

---

3 This capability was added in Python 3.7.

cated situations, such as reentrant concurrent code, and this extra functionality comes with a performance penalty. You should only use `queue.SimpleQueue()` if you need concurrent access to the queue.

The `deque` is also the fastest implementation here. It is specially coded for raw speed and is the queue implementation you should choose if speed is critical. Table 8-3 shows that the `deque` implementation offers the best implementation; it also shows how `list` provides O(N) runtime performance for dequeuing a value. Once again, you must avoid using a generic `list` as a queue, since all other queue implementations offer O(1) runtime performance.

Table 8-3. Queue runtime performance comparisons when dequeuing a value

N	list	deque	SimpleQueue	Queue
1,024	0.012	0.004	0.114	0.005
2,048	0.021	0.004	0.115	0.005
4,096	0.043	0.004	0.115	0.005
8,192	0.095	0.004	0.115	0.005
16,384	0.187	0.004	0.115	0.005

The essential data types provided by Python are flexible enough to be used in a variety of ways, but you need to make sure you choose the appropriate data structures to realize the most efficient code.

Heap and Priority Queue Implementations

Python provides a `heapq` module providing a min binary heap, as described in Chapter 4. Instead of using the 1-based indexing strategy I used in Chapter 4, this module uses 0-based indexing.

The `heapq.heapify(h)` function constructs a heap from a list, `h`, containing the initial values to place in the heap. Alternatively, simply set `h` to the empty list `[]`, and invoke `heapq.heappush(h, value)` to add `value` to the heap. To remove the smallest value from the heap, use the `heapq.heappop(h)` function. This heap implementation has two specialized functions:

`heapq.heappushpop(h, value)`
>   Adds `value` onto the heap and then removes and returns the smallest value from heap

`heapq.heapreplace(h, value)`
>   Removes and returns the smallest value from the heap and also adds `value` onto the heap

These functions are all applied directly to the parameter h, making it easy to integrate with your existing code. The contents of h reflect the array-based storage of a heap, as described in Chapter 4.

The `queue.PriorityQueue(maxSize=0)` method (*https://oreil.ly/sUiZd*) constructs and returns a min priority queue whose entries are added using `put(item)` function calls, where `item` is the tuple (`priority, value`). Retrieve the value with lowest priority using the `get()` function.

There is no built-in indexed min priority queue, which is not surprising because this data type is typically only needed for specialized graph algorithms, such as Dijkstra's algorithm (presented in Chapter 7). The `IndexedMinPQ` class I developed shows how to compose different data structures together to achieve an efficient `decrease_prior ity()` function.

Future Exploration

This book has only scratched the surface of the incredibly rich field of algorithms. There are numerous avenues you can explore, including different application domains and algorithmic approaches:

Computational geometry
   Numerous real-world problems involve data sets with two-dimensional points, or even higher dimensions. Within this application domain, there are many algorithms that build from the standard techniques I've introduced—such as divide and conquer—and introduce their own data structures to efficiently solve problems. The most popular data structures include *k*-d trees, Quadtrees (for partitioning two-dimensional spaces), Octrees (for partitioning three-dimensional spaces), and R-trees for indexing multidimensional data sets. As you can see, the essential concept of a binary tree has been explored repeatedly in different application domains.

Dynamic programming
   Floyd–Warshall is an example of dynamic programming applied to solving the single-source shortest path problem. There are many other algorithms that take advantage of dynamic programming. You can learn more about this technique in my *Algorithms in a Nutshell* (*https://oreil.ly/1lXRF*) book also published by O'Reilly.

Parallel and distributed algorithms
   The algorithms presented in this book are essentially single-threaded and execute on a single computer to produce a result. Often a problem can be decomposed into multiple constituent parts that can be executed independently, in parallel.

The control structures become more complicated, but impressive speed-up can be achieved through parallelism.

Approximation algorithms

In many real-worlds scenarios, you might be satisfied with an algorithm that efficiently computes an approximate answer to a really challenging problem. These algorithms are noticeably less computationally expensive than those producing the exact answer.

Probabilistic algorithms

Instead of producing the exact same result when given the exact same input, a probabilistic algorithm introduces randomness into its logic, and it produces different results when given the same input. By running these less complex algorithms a large number of times, the average overall runs can converge on the actual result, which would otherwise be prohibitively expensive to compute.

No single book can cover the breadth of the field of algorithms. In 1962, Donald Knuth, a towering figure in computer science, began writing a 12-chapter book, *The Art of Computer Programming* (Addison-Wesley). Today—59 years later—the project has produced three volumes (published in 1968, 1969, and 1973) and the first part of volume 4 (published in 2011). With three more volumes planned for publication, the project is still not complete!

You can find countless ways to continue your study of algorithms, and I hope you are inspired to use this knowledge to improve the performance of your software applications.

Index

Symbols

% (modulo operator), 61
& (bitwise and), 64, 252
// (integer division), 134
< (less-than operator), 4
> (greater-than operator), 15, 118
_insert() method, 165, 182
_remove() method, 182
_remove_max() function, 188
_remove_min() method, 182
Θ (theta), 53
Ω (omega), 53

A

active search space, 208, 211, 216, 218, 220
additive constant, 36
adjacency list, 213
adjacency matrix, 213
algorithm
    analysis, 36
      (see also asymptotic analysis)
    complexity, 23
      (see also space complexity, time complexity)
    definition of, 2
    performance, 32, 148
      (see also performance classes, performance prediction)
all-pairs shortest path, 237-240
alpha, 85
alternate() algorithm, 7-11
amortized analysis
    amortized constant O(1), 227
    amortized O(1), 57, 249, 251, 251

amortized performance, 118
approximation algorithms, 257
array data structure, 4
    for heap storage, 113
    for open addressing, 66
    geometric resizing, 57, 81, 85, 118, 175, 249
Art of Computer Programming, The, 257
ASCII, 61
asymptotic analysis, 36-39
AVL property, 176
AVL tree, 176-179, 190, 193

B

bag, 249
base case, 132, 155
Bellman–Ford algorithm, 234-237
best case problem instance, 8, 39
Big O notation, 38
Bignum structure, 33
Binary Array Search, 101, 159
Binary Array Search algorithm, 42-49
binary heap, 104-112
    levels, 105-106
    max binary heap, 97, 104-106
    min binary heap, 97, 106, 118, 255
    sink() function, 110-112, 116-118
    swim() function, 107-109, 114-116
binary search tree
    about, 159-165
    performance, 174-176
    priority queue as, 187-190
    self-balancing, 176-185
    symbol table as, 185-187
    traversing, 171-172

(see also descendants, leaf node, parent
    nodes, root nodes)
numpy, 31

O

open addressing, 66, 77, 79, 82
order of a function, 38

P

palindromes, 25
parallel algorithms, 256
parent nodes, 164
path, 107, 198
perfect hashing, 86-89
perfect_hash() function, 87
performance classes, 34-36
performance comparison, 148
performance prediction, 7-12, 30-32
Peters, Tim, 148
polynomial complexity class, 50
postorder traversal, 172
prefix order, 193
preorder traversal, 172
priority, 97
priority queue, 97-102, 98, 187, 250
    (see also heap-based priority queue)
probabilistic algorithms, 257
problem instance, 2, 8, 39
    (see also best case problem instance, worst
        case problem instance)
programming effort, 13
project, 196
put() function, 67, 70, 74, 85
put(k, v) function, 186
Python, 4, 11, 33, 60, 62, 148, 159, 251-256
    enumerate, 87
    generators, 40, 89, 171
    interpreter, 3
    itertools, 54
    NetworkX, 196, 199, 245
    NumPy, 31
    perfect-hash, 87, 91
    Python 2, 2, 40
    Python 3, 3, 40
    range, 6, 40, 41, 89
    RuntimeError, 65, 68, 203, 228, 246
    SciPy, 31, 52
    sys, 40, 41
    ValueError, 6

__contains()__, 166, 186
__iter()__, 89, 94, 171, 172, 186
Python-2, 64
Python-3, 64

Q

quadratic complexity class, 39
quadratic models, 31-32, 34
quadratic polynomial, 31
quadratic sorting algorithm, 127-129
queue, 98, 249
    (see also circular queue, dequeue operation,
        enqueue operation, priority queue)
quicksort, 141-145, 148

R

RAM (Random Access Memory), 24
recursion, 131-134, 154
    (see also binary tree recursive data struc-
        ture)
recursive algorithm, 132
recursive case, 132, 155
recursive data structure, 154
    (see also binary tree recursive data struc-
        ture, linked lists)
recursive helper function, 163
references (see links)
relax() function, 233
relaxing an edge, 231
remove(k) function, 76
root nodes, 160, 174
rotate left, 179, 193
rotate left-right, 180, 193
rotate right, 179, 193
rotate right-left, 193

S

search
    binary tree, 165-166
    linear probing, 71-74
    open addressing, 66-68
    ordered array, 42-48
selection sort, 125-127, 129-131
separate chaining technique, 71
set data type, 253
simple graph, 198
simple uniform hashing, 85
sink() method, 117

About the Author

George Heineman is a professor of computer science with over 20 years of experience in software engineering and algorithms. He is the author of *Algorithms in a Nutshell* (2nd edition) and numerous O'Reilly live training offerings, including "Exploring Algorithms in Python" and "Working with Algorithms in Python." He has a lifelong interest in logical and mathematical puzzles. He is the inventor of Sujiken® puzzles, a variation of Sudoku, and Trexagon puzzles.

Colophon

The animal on the cover of *Learning Algorithms* is a Chesapeake blue crab (*Callinectes sapidus*). The genus name *Callinectes* comes from the Greek for "beautiful swimmer" and the species name *sapidus* is Latin for "savory." The crab's color is produced by pigments in the shell, including alpha-crustacyanin, which interacts with the red pigment astaxanthin to form a green-blue color. When a crab is cooked, the alpha-crustacyanin breaks down and the crab's shell turns a bright orange-red color.

The blue crab is native to the western edge of the Atlantic Ocean and to Gulf of Mexico. It was introduced to Japanese and European waters through water ballast as far back as 1901. Recently, it's thought that their habitat is expanding due to warming waters from climate change.

Crab eggs hatch in coastal waters and are carried into deeper waters by tides. The larvae go through eight planktonic stages before reaching the juvenile phase, when they appear similar to adults. They grow by molting, the process of shedding the exoskeleton to expose a new, larger exoskeleton. It's thought that for blue crabs the number of molts in a lifetime is fixed at approximately 25. They can grow to a width of about 9 inches. Males have a slender abdomen and females have a wide, rounded abdomen. Males and females also have subtle differences in coloration.

Many of the animals on O'Reilly covers are endangered; all of them are important to the world.

The cover illustration is by Karen Montgomery, based on a black and white engraving from *Animal Life in the Sea and on the Land*. The cover fonts are Gilroy Semibold and Guardian Sans. The text font is Adobe Minion Pro; the heading font is Adobe Myriad Condensed; and the code font is Dalton Maag's Ubuntu Mono.

O'REILLY®

There's much more where this came from.

Experience books, videos, live online training courses, and more from O'Reilly and our 200+ partners—all in one place.

Learn more at oreilly.com/online-learning